ROBERT W. MURRAY

GENERAL ZOOLOGY LABORATORY GUIDE
Third Edition

JAMES C. UNDERHILL
MAGNUS OLSON
WILLIAM D. SCHMID
DAVID J. MERRELL

University of Minnesota
Minneapolis - St. Paul

Burgess Publishing Company
Minneapolis, Minnesota

COVER: Male chorus frog, *Pseudacris nigrita,* calling in early spring. Photograph by William D. Schmid.

Copyright © 1978, 1974, 1972 by Burgess Publishing Company
Printed in the United States of America
ISBN 0-8087-2108-9

All rights reserved. No part of this book may be reproduced in any form whatsoever, by photograph or mimeograph or by any other means, by broadcast or transmission, by translation into any kind of language, nor by recording electronically or otherwise, without permission in writing from the publisher, except by a reviewer, who may quote brief passages in critical articles and reviews.

0 9 8 7   5 4 3 2

CONTENTS

Preface v

Acknowledgments v

Introduction vi

1. The Compound Microscope 1
   I. Parts of the Microscope.   II. Cautions to Observe.   III. Using the Microscope.

2. Cells 6
   I. Cell Structure.

3. Mitosis and Meiosis 8
   I. Mitosis and Somatic Cell Division.   II. Meiosis.

4. Phylum Protozoa 14
   I. Class Mastigophora.   II. Class Sarcodina.   III. Class Ciliata.
   IV. Sporozoans.

5. Introduction to Metazoa: Phylum Porifera 19
   I. Metazoa.   II. Sponges.

6. Phylum Coelenterata 22
   I. *Obelia.*   II. *Gonionemus.*   III. *Hydra.*   IV. Other Coelenterates.
   V. Interpretation of Coelenterate Life Cycles.

7. Phylum Platyhelminthes 29
   I. Class Turbellaria.   II. Class Trematoda.   III. Class Cestoda.

8. Phylum Nematoda 36
   I. *Ascaris.*   II. Vinegar Eel.   III. Other Nematodes and Rotifers.

9. Phylum Annelida 42
   I. *Nereis.*   II. Earthworm.

10. Phylum Mollusca 50
   I. Freshwater Clam.   II. Other Molluscs.

11. Phylum Arthropoda 54
   I. Crayfish.   II. Grasshopper.   III. Insect Metamorphosis.
   IV. Orders of Insects.   V. Other Arthropods.   VI. Phylum Onychophora.

12. Phylum Echinodermata 70
   I. Starfish.   II. Other Echinoderms.   III. Embryology and Development.
   IV. Relation of Echinoderms to Chordates.

13. Phylum Chordata 74
I. Subphylum Urochordata. II. Subphylum Cephalochordata.
III. Relation to Vertebrates. IV. Subphylum Vertebrata.
V. Animal Communication: Sound (Vocal → Auditory) Stimulation.

14. Fetal Pig 81
I. General External Anatomy. II. Skeletal System.
III. General Internal Anatomy. IV. Circulatory System.
V. Heart and Pulmonary Circulation. VI. Respiratory System and its Operation.
VII. Digestive System. VIII. Excretory System.
IX. Reproductive System. X. Nervous System. XI. Cells and Tissues.
XII. Whole Blood: Hematocrit and Hemoglobin Content.

15. Taxonomy of Mammals 117

16. Human Genetics 120

PREFACE

This laboratory manual, now in its third edition, is designed for a 1-term course in General Zoology. It is designed to provide a survey of the animal kingdom dealing with the major phyla and the study of the anatomy of a mammal, specifically, the fetal pig. The exercises are designed to provide a minimum of work and allow the instructor to expand on them through demonstrations, movies, additional questions, and quizzes.

Although the manual is expanded, the format of concise directions and detailed illustrations remains the same. We have added exercises on Porifera, taxonomy of mammal skulls, and human genetics. The section on the fetal pig, which is the most intensive and sustained exercise in the manual, has been reorganized and, we hope, improved in response to the requests of other instructors who use the manual. We have brought together the sections on cells and tissues and the properties of blood, separating them from the dissection. We have added a section on the skeleton, enlarged on the discussion of patterns of blood circulation in fetal and postnatal placental mammals, and added five illustrations.

The manual has been in use at the University of Minnesota for 7 years. We wish to express our sincere appreciation to all students, teaching assistants, and colleagues who have contributed to its improvement with their suggestions and comments.

ACKNOWLEDGMENTS

Exercises from the *General Biology Laboratory Guide* by Magnus Olson, Norman S. Kerr, and Kenneth S. Skjegstad have been included herein with some modifications. Final preparation of drawings for publication was done by Alice Benjamin, Nancy Arko, Marilyn Steere, and Magnus Olson. We wish to thank Dr. Louise Rollins for the use of photographs of the sponge spicules and the gemmule. We especially appreciate the assistance, important suggestions, and comments of Dr. Gary Phillips, who guided this revision to publication.

INTRODUCTION

USE OF THE LABORATORY MANUAL. This manual was prepared to facilitate your work in the laboratory. It contains information, directions, and suggestions to aid you in your work. The most important items are in **bold type** for emphasis and merit your special attention. The first of these is that you must **read the pertinent exercise and text discussion in advance** of the laboratory period in which the exercise is scheduled. Come to your laboratory section prepared to begin the exercise.

THE LABORATORY KIT. Besides this manual you must bring the following items to the laboratory:

1 drawing pencil (3H)
rubber eraser
15-cm ruler
scalpel
2 dissecting needles
forceps
scissors

YOUR INSTRUCTOR is in the laboratory to help you. **Do not hesitate to ask questions.** A major objective of this course is to stimulate your curiosity about animals. Do not be afraid that asking many questions will merely display your ignorance; it takes intelligence to ask questions. You will learn more from the laboratory exercises if you take advantage of your opportunities to discuss the subjects under consideration with the laboratory instructor. However, you are not expected to ask a question which would have been answered by reading the appropriate sections from the manual and text beforehand. Depend on your own resources where possible.

The manual is both a guide and a workbook. Space is provided for laboratory notes, drawings, and answers to specific questions. You will not be asked to hand in laboratory reports, themes, or drawings. However, when the manual explicitly calls for you to record information, to prepare a drawing, or to write out the answers to questions, these should be done in the blank spaces provided. From time to time your laboratory instructor will inspect your manual, and when appropriate he or she will discuss and criticize what you have drawn and written. It is thus essential that you write legibly and draw neatly.

You should take notes and make quick sketches for use in review even where no specific instructions are given.

DRAWINGS. Where a drawing is required, do not make elaborate efforts to portray what can be shown simply. Artistic ability is unessential (fortunately for most of us); you need only the ability and patience to draw clear outlines showing the shape of an object and its parts. Use a No. 3H pencil. Make each line distinct. Do not shade. Make drawings large enough so that the smallest detail you wish to represent will be clear. **Always label the drawing neatly.** Write carefully or print.

1. THE COMPOUND MICROSCOPE

Many animals are large enough to be studied with the unaided eye, and you can learn much by looking at them. However, some animals and certain parts of all animals are much too small to be studied without the aid of a **microscope**. A considerable portion of your laboratory work will require the use of this instrument. Therefore, you must learn what a microscope is, how it functions, how to use it to best advantage, and how to care for it properly. A microscope is a delicate precision instrument. If properly handled, it will give good service for many years; if carelessly treated, it can suffer $100 damage in a few seconds. Carefully read the following description of the instrument and the instructions and precautions about its use. Special instructions will be given by your laboratory instructor.

I. PARTS OF THE MICROSCOPE

When picking up the microscope **always** use **two hands**, one grasping the **arm** (Figure 1-1) and the other supporting the **base**. Always rest the microscope directly on the laboratory table, not on a notebook or other less reliable support. Keep the microscope squarely in front of you when it is in use. Avoid leaving it in a precarious position near the edge of the desk. When you have completed your work return the microscope to the cabinet.

With the aid of Figure 1-1, identify the parts of your microscope and learn the function of each. The base and arm were mentioned previously. The object to be examined is mounted on a slide that is placed upon the **stage**. Light passes from the **illuminator** through the slide and into the **objective lens** (mounted on a revolving **nosepiece**). The light is focused by the objective lens to form an image by means of prisms in the **barrel**. The top lens of the microscope, called the **ocular** (ocular lens), magnifies the image as the light passes on to the eye. The **diaphragm** (iris diaphragm) mounted beneath the stage is used to control the amount of light passing through the lens. Some microscopes have a **substage condenser** associated with the diaphragm which concentrates the light on the specimen. Two focusing adjustments raise and lower the stage; these are a **coarse adjustment knob**, the large knurled knob, and a **fine adjustment knob**, the small knurled knob.

The revolving **nosepiece** supports a **low-** and a **high-power objective**. These may be marked in several ways. The low-power objective may have 10X, 16 mm, N.A. .25, or all three engraved on it and may also have green banding. The 10X means that the lens's contribution to the microscope's magnifying power is tenfold; that is, it forms an image inside the barrel (or body tube, Figure 1-2) that is 10 times as large in any dimension as the specimen on the slide. The 16 mm is not in itself a measure of magnification, but instead indicates the distance between the specimen and the focal point of the objective lens. The N.A. is an abbreviation for numerical aperture. The high-power objective is marked 43X, 4 mm, N.A. .55, or all three and has yellow banding.

Some microscopes have an extra-low-power 3X objective. The microscopes used for demonstrations are similar to the microscope in Figure 1-2 and may have three or four objective lenses rather than two. One may be a high-dry lens with a magnification between 10X and 43X, and another a 100X oil-immersion objective. The latter objective lens provides the maximum resolution possible.

The ocular lens may also be marked with its magnification power. The usual ocular is 7.5X or 10X. The significance of this mark is the same as that on the objective; it indicates the contribution of this lens to the overall magnification. Using a 10X ocular and a 10X objective, the overall magnification will be the product of these two numbers, 100. Using a 10X ocular and the 43X objective, the final magnification is 430.

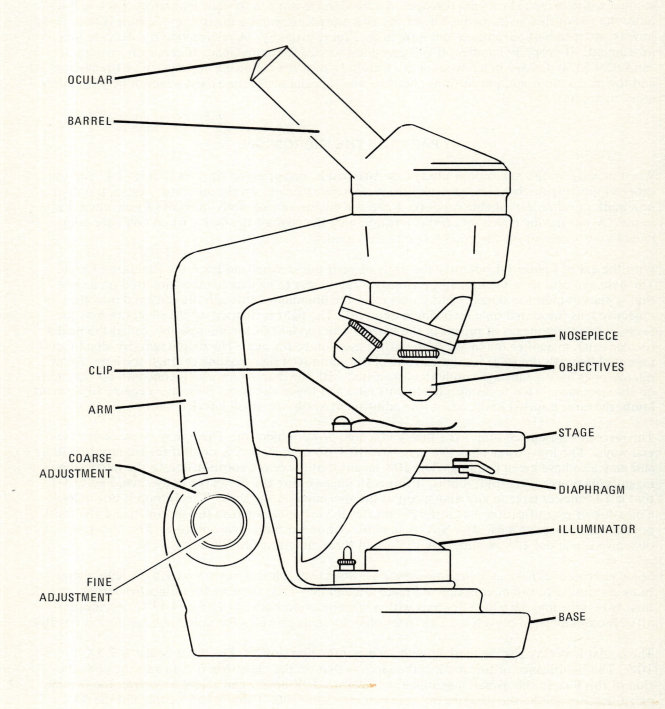

FIGURE 1-1 THE COMPOUND MICROSCOPE

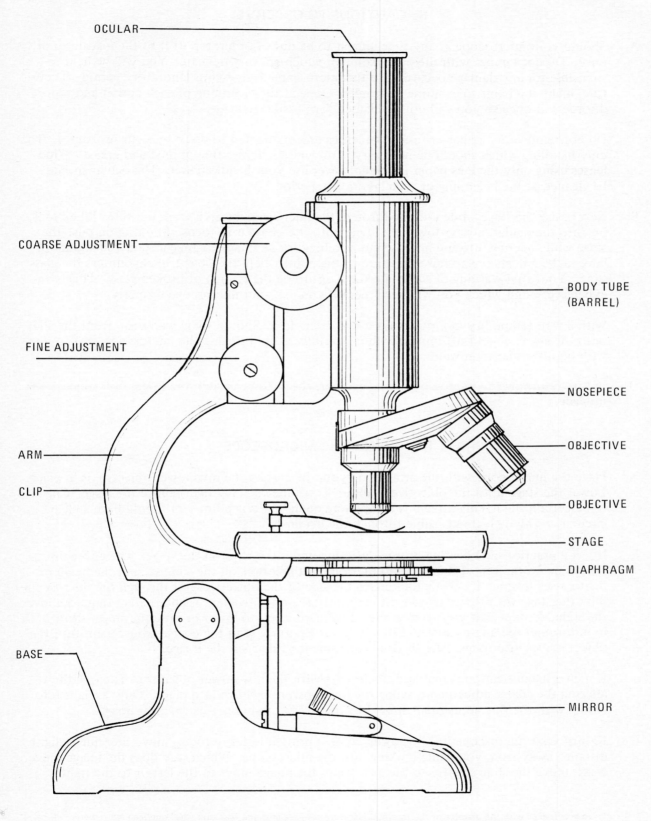

FIGURE 1-2 THE COMPOUND MICROSCOPE USED FOR DEMONSTRATION

II. CAUTIONS TO OBSERVE

A. Should your microscope at any time appear to be out of order, report it to the instructor at once. Do not tamper with the parts, some of which are very delicate. You will be held responsible for any damage occurring to the microscope during your laboratory period. Therefore, make it a habit to examine your microscope at the beginning of each period and report disorders at once so you will not be blamed for possible damage.

Do not remove the ocular or touch the lenses unless directed to do so by your instructor. The only handling which should be necessary is to wipe the lenses free of dust and grease. **Wipe lenses using only the lens paper provided; never use your handkerchief.** The ocular should be cleaned at the beginning of each laboratory period.

B. Never raise the stage while looking through the microscope; watch from the side. When looking into the ocular, always lower the stage until the image is in focus. Should you raise the stage while peering into the microscope watching for an image to appear, you will sooner or later make a mistake and raise the stage to the point where the objective encounters the glass slide. Your first warning of such an error is apt to be the crunch of broken glass. This is an expensive sound which you will never hear if you use your microscope properly.

C. With a wet, temporary mount always use a cover glass. Should some water run from the slide onto the stage, wipe it off immediately. Carefully avoid bringing the microscope into contact with liquids other than water.

D. Before returning the microscope to its cabinet, check to be certain the low-power objective is in place.

III. USING THE MICROSCOPE

A. Place the microscope with the arm facing you. Be certain the low-power objective is in place. Lower the stage until the objective is about 1 centimetre[1] (1/2 inch) above the stage. Check to be certain the iris diaphragm is completely open. At this point you should have a clear, circular, evenly light **field** visible through the ocular.

B. Insert a **practice slide bearing printed lettering** under the clips on the stage. The cover slip should be on the top of the slide. Center one of the letters on the slide so that the beam of light passes through it. While looking to the side of the objective (**not through the microscope**), raise the stage with the coarse adjustment until the objective lens is about 1 centimetre above the slide. Note which way to turn the adjustment knob to lower or raise the stage. **Now look through the microscope** and slowly lower the stage with the coarse adjustment until the object comes into focus. Use the fine adjustment to sharpen the image.

If higher magnification is required, focus first with the low-power objective. Then, **without altering the coarse adjustment**, swing the high-power objective into place. Only a slight refocusing with the fine adjustment should be necessary to obtain a sharp, clear image.

C. Return your microscope to low power. With a printed letter in focus, move the slide a slight distance away from you while looking into the microscope. Which way does the image move? Next, move the slide slightly to the left. Does the image move to the left or to the right?

---

1. In accordance with the practice of the International Bureau of Weights and Measures, the American word *meter* will be spelled *metre.*

D. Examine a slide of **colored threads**. Using a high power, focus up and down with the fine adjustment to determine which color thread is on top, which is in the middle, and which is on the bottom. Record your results and ask your instructor to check them for you. Can you determine all three dimensions of an object with a microscope?

E. You can determine the approximate size of an object viewed through the microscope in the following manner. Make a wet mount of a small piece of millimetre-squared graph paper. Focus on one of the squares with the low-power objective in place. What is the diameter of the low-power microscope field? Next, focus with high power on the graph paper. What is the diameter of the high-power field? With this information you can estimate the approximate size of other objects viewed through the microscope. Whenever you use a different microscope, it will be necessary to calibrate the new microscope in a similar fashion.

F. **Examining a temporary slide.** It is often desirable to observe something under the microscope without making a permanent mount. Living objects are always studied in this way. The object to be examined is mounted in a drop of water and covered with a glass **cover slip**.

Transfer a small drop of **pond water** to a clean slide. Following your instructor's directions, add a cover slip. Examine the slide under low power. It may be necessary to reduce the light intensity by partially closing the iris diaphragm. You should see an abundance of microscopic plants and animals in the drop of water. Can you distinguish living organisms from nonliving structures? How many kinds of living things can you find on your slide? Examine some of the organisms under high power, noting the intricate structures that can be observed.

By noting the time it takes an organism to swim across the microscope field, estimate its speed of swimming.

QUESTIONS

1. What magnifying power do you expect from a microscope having a 95× objective and a 7.5× ocular?

2. Why is your microscope referred to as a **compound** microscope?

3. What is a simple microscope?

2. CELLS

The cell theory, one of the great foundations of modern biology, states that the **cell** is the basic structural and functional unit of living organisms. In this exercise you will examine a few of the many types of cells which constitute the "building blocks" for all living organisms. As you observe various types of cells, ask yourself the following questions. What is the function of each structural element of the cell? Which structures are common to all cells? Which are found in animal cells? In addition to studying living cells, you will learn some simple cytological techniques (**cytology** = the study of cells) such as staining. This exercise will be more meaningful if you read pertinent sections of your textbook before coming into the laboratory.

In this and later exercises you will examine prepared slides (permanent mounts) in which the object on the slide has been treated in various ways which make certain structures readily visible. It may be helpful to insert a few remarks concerning these permanent slides. The object to be studied is killed by a chemical agent which does not damage the structure but rather preserves it. The object is then embedded in paraffin, and the block of paraffin is cut into very thin sections with a machine called a **microtome**. The sections are mounted on slides and run through a series of solutions and dyes which stain various structures differently. Then the permanent mount is made by adding a drop of resinous fluid (balsam) in a solvent that will evaporate. A cover glass is added and, when the balsam is hard (solvent removed), the object is permanently available for microscopic examination. **Observe the demonstrations of this technique.**

I. CELL STRUCTURE

A. Starfish Immature Egg

Examine the prepared slide labeled **starfish immature ovum** under low power. The unfertilized eggs have been stained with any of a number of dyes before being mounted. Center one such cell in the low-power field of your microscope. Note the two most obvious parts of the cell: the central **nucleus** and the surrounding **cytoplasm**. The outermost layer of the cytoplasm is the **cell membrane**. The nucleus is separated from the cytoplasm by a **nuclear membrane**. Within the nucleus note the **chromatin granules** and the rather prominent **nucleolus**. What is the approximate diameter of the ovum?

Draw one starfish egg in the space provided and neatly label the structures that you have observed. Get your laboratory instructor's approval of your drawing.

Starfish Immature Ovum, _____X

B. **Epithelial Cells**

   1. SQUAMOUS EPITHELIUM

      Gently and carefully scrape the inner surface of your cheek with a toothpick. Place the substance you obtain on a clean slide and add a drop of saliva. Spread it thoroughly with a dissecting needle and add a cover slip. The **squame cells** are very transparent, so it will be necessary to lower the light intensity in your microscope.

      Identify the plasma membrane which bounds the cell, the granular **cytoplasm**, and the rather clear **nucleus**. After examining the cells unstained, stain them, following your laboratory instructor's directions. Are the structures now more clearly visible? Make a careful drawing of several squame cells in the space provided.

      Squame Cells, _____ X

C. **The Nucleus**

   The hereditary factors of the nucleus, called **genes**, are located on bodies called **chromosomes** (= colored bodies) because they stain with certain dyes. Chromosomes are usually visible only in nuclei that are undergoing division. Examine the demonstrations of *Drosophila* salivary gland giant chromosomes.

D. **The Electron Microscope**

   The wavelength of visible light determines the maximum resolution possible with the light microscope. Even with the best research microscopes, objects smaller than 0.1 micrometre (μm) cannot be resolved. The electron microscope substitutes a beam of electrons for the beam of light. Because the electron beam operates in a vacuum, only fixed, sectioned objects may be used. The results obtained are recorded on photographic film, from which prints are made in the usual way.

   Examine the electron micrographs that are on demonstration. Note the wealth of structure in the cell too fine to be seen with the light microscope.

3. MITOSIS AND MEIOSIS

I. MITOSIS AND SOMATIC CELL DIVISION

For most of you this exercise will be a review but, because of the importance of this biological process, you are urged to study your slide and demonstrations thoroughly. In your own development from a single cell to a complex aggregation of billions of cells, the complex series of events in **mitosis** occurred at each cell division. You will recall that although the process of mitosis and cell division is a continuous one, it has been divided into stages or phases for convenience in study. You will study the skin of the salamander tadpole tail, a fast-growing tissue showing many mitoses. Try to identify all of the stages you find on your slide.

The conventional phases and the events correlated with each are shown in Figure 3-1 and are described as follows.

A. **Interphase**

In this stage the chromosomes are enormously expanded and are not visible as such. It is during the interphase that the DNA is replicated, a condition that will become obvious in the next phase. Note the chromatin granules, nucleoli, and nuclear membrane.

B. **Prophase**

The chromosomes become visible at first as they condense and coil as irregular rough threads. Gradually the nuclear membrane and nucleoli disappear.

In the middle prophase the chromosomes have condensed into darkly staining, long, smooth, tangled threads. In the late prophase they have condensed still further and have now assumed the form characteristic of the species (mostly V-shaped in the salamander). Furthermore, it is now possible to see the double nature of the chromosomes quite readily. Each of the parts is known as a **chromatid**. The two chromatids are joined at the apex of the V by a centromere (not visible on your slides).

C. **Metaphase**

Chromosomes become arranged in an **equatorial plate**. Not visible on your slides but shown on demonstration are the **spindle fibers** that run from the **centrioles** at the poles and attach to the chromosomes.

D. **Anaphase**

The two chromatids of each pair separate and are drawn to opposite poles of the cell.

E. **Telophase**

Chromatids complete their movement to the poles and begin their uncoiling and expansion. In Figure 3-1 the chromosomes again are assuming an irregular, rough outline. Cytoplasmic division begins in advanced telophase.

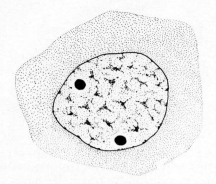
INTERPHASE

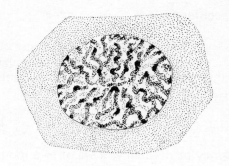
EARLY PROPHASE

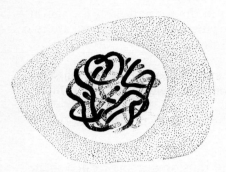
MIDDLE PROPHASE

LATE PROPHASE

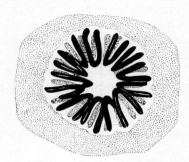
METAPHASE (POLAR VIEW)

METAPHASE (SIDE VIEW)

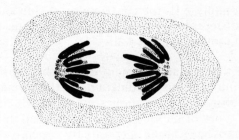
ANAPHASE

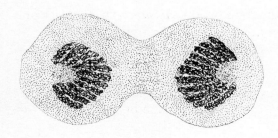
TELOPHASE

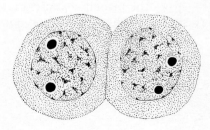
DAUGHTER CELLS

FIGURE 3-1 MITOSIS

F. Daughter Cells

Cytoplasmic division has been completed and the chromosomes are completely expanded. The nucleoli and nuclear membrane have reappeared. Two daughter cells, somewhat smaller and more darkly staining than the parent cell, have formed.

II. MEIOSIS

Meiosis may be defined as a set of two modified mitoses during which the chromosomes divide only once. As a result of meiosis the number of chromosomes is reduced by one-half. Thus, when the egg and the sperm cell fuse, the normal double or **diploid** number is restored, and the chromosome number remains constant from generation to generation.

Figures 3-2 and 3-3 are diagrams of the meiotic divisions that occur in the formation of the sperm (**spermatogenesis**) and the formation and maturation of the egg (**oogenesis**). Because there are two divisions, the various stages of the two divisions are designated by I or II.

In prophase I, homologous pairs of chromosomes line up with one another lengthwise (**synapsis**). Each chromosome is visibly double except for one point of attachment known as the **centromere**.

Thus, on the equatorial plate at metaphase I there are four-parted bodies known as **tetrads**. Each of the four parts is known as a **chromatid**.

In anaphase I, the two chromatids attached at each centromere (and thus originating from a single chromosome) move to opposite poles of the cell. Homologous pairs of chromosomes have thus been separated into daughter cells.

In metaphase II, the centromeres divide and, thus, in anaphase II each chromatid of the pair moves to a different pole of the cell. The net result of these two divisions is the reduction of the chromosome number by one-half.

Note that the arrangement of the chromosomes on the equatorial plate is purely a matter of chance as is indicated in the diagrams of the primary spermatocyte and the oocyte in Figures 3-2 and 3-3. Therefore, in the examples shown in which there are four chromosomes (two homologous pairs), four types of haploid nuclei are formed in equal numbers. This variation in the arrangement of chromosomes alone accounts for much of the diversity that arises during sexual reproduction. The genes on opposite members of homologous chromosomes, remember, have similar functions but are not necessarily identical. The human with 46 chromosomes (23 homologous pairs) will produce 2^{23} or 8,388,608 different types of gametes.

Figure 3-4 shows the principal stages in oogenesis in the roundworm, *Ascaris*. These stages or stages very similar to them will be on demonstration.

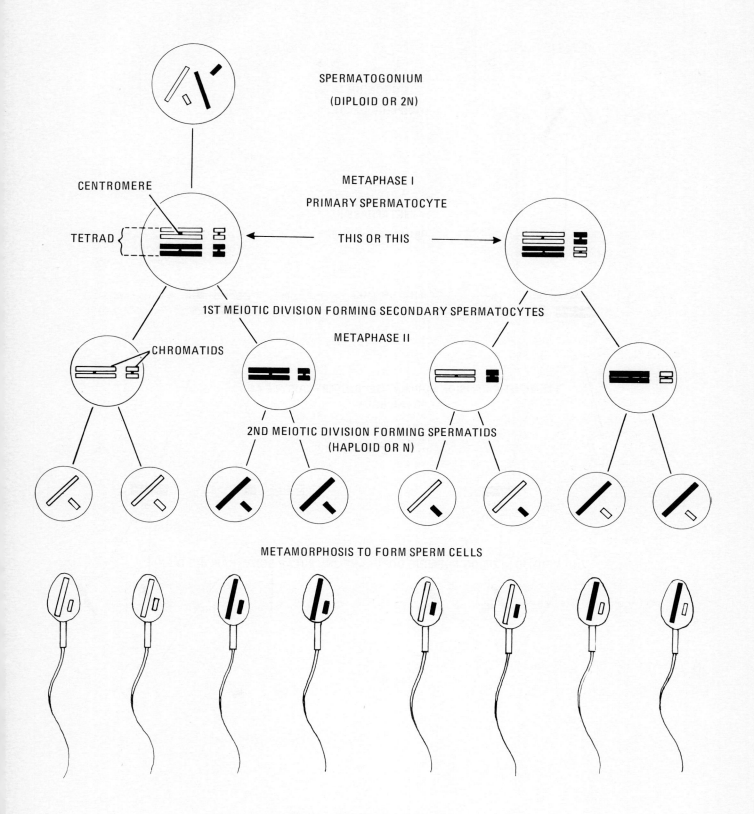

FIGURE 3-2 SPERMATOGENESIS

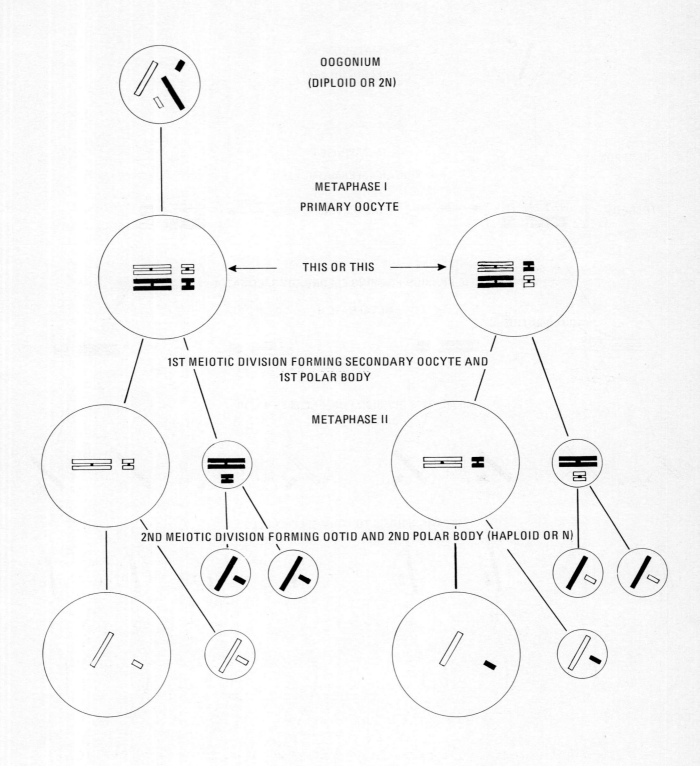

FIGURE 3-3 OOGENESIS

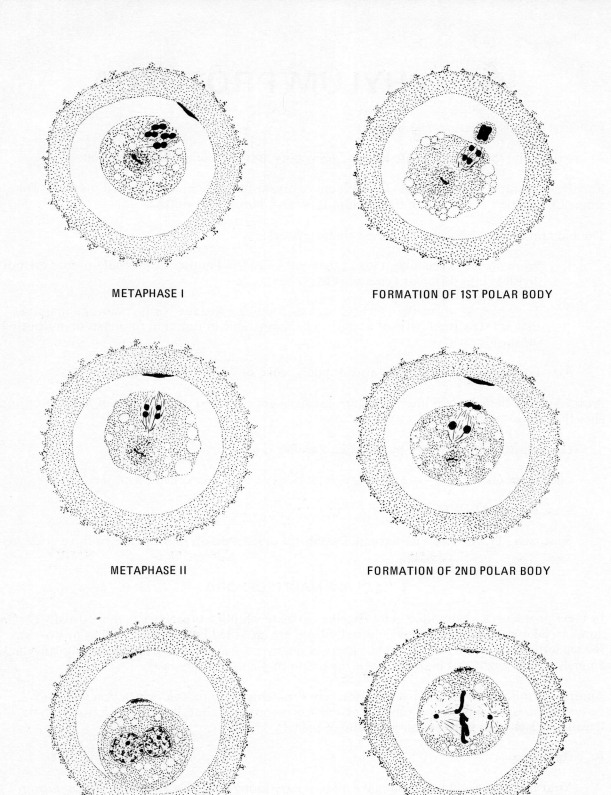

FIGURE 3-4 MEIOSIS IN *ASCARIS* EGG

4. PHYLUM PROTOZOA

Protozoa (protos = first, zoon = animal).

Approximately 25,000 species are known, depending on the system of classification.

Relatively primitive, single-celled (some colonial) organisms are sometimes called protists. The animallike protists are placed in the phylum Protozoa, kingdom Animalia.

The phylum Protozoa has the following characteristics.

1. Protozoans are **acellular**, that is, they are specialized masses of protoplasm that are not divisible into different types of cells.

2. They possess **organelles** (such as nucleus, feeding apparatus, or locomotor apparatus) that are structural parts of a single cell, comparable in function to organs in multicellular organisms.

3. Locomotion is by flagella, pseudopodia, cilia, or direct cell movement.

Protozoans are here divided into four large, artificial groups based on the types of locomotor organelles they have:

Class Mastigophora. Move by whiplike flagella. *[Euglena, Peranema]*

Class Sarcodina. Move by temporary extensions of the body called pseudopodia. *[Amoeba, Difflugia, Arcella]*

Class Ciliata. Move by cilia. *[Paramecium, Vorticella, Euplotes]*

Sporozoa. Protozoans that have no locomotor organelles; often divided into three classes. *[Plasmodium vivax – No locomotion – All Parasites]*

I. CLASS MASTIGOPHORA

In these protozoans locomotion is by **flagella**, elongate whiplike organelles that are usually few in number. Many flagellates possess chlorophyll and are classified both as algae and as protozoans. No separation is possible, or desirable. There is simply no sharp dividing line between plants and animals. What might this indicate about the origin of plants and animals?

Among the unicellular flagellated organisms are *Peranema* and *Euglena*, which you will study. *Peranema* does not have chlorophyll. *Euglena* has chlorophyll, but, like other photosynthetic flagellates, lacks cell walls (a characteristic of plants).

A. <u>Peranema</u>

Your instructor will help you make a temporary mount of living *Peranema trichophorum* (Figure 4-1). *Peranema* feeds as a **saprophyte** by absorbing nutrients through its plasma membrane or as a **phagotroph** by engulfing other microorganisms.

Examine your preparation of *Peranema*, first using low power and then high power. By lowering the intensity of illumination, you should readily see its flagellum. Only the distal tip of the flagellum beats, making it easy to see. Note that the cytoplasm is filled with

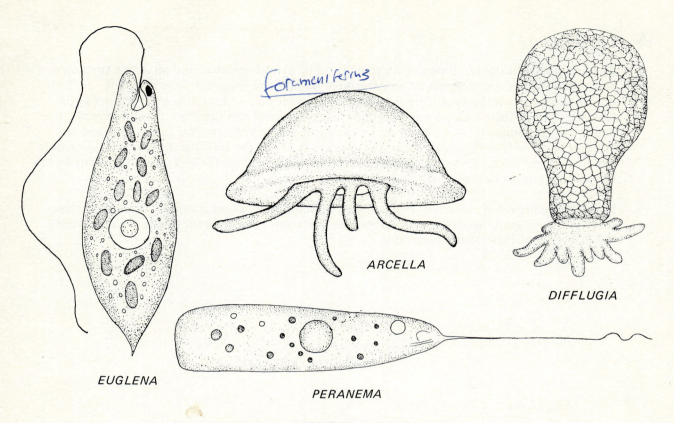

FIGURE 4-1 REPRESENTATIVE PROTOZOANS

vacuoles, granules, and droplets of various types. What might be the composition of some of these structures? Describe the locomotion of *Peranema*.

B. **Euglena**

Euglena is an example of a photosynthetic flagellate (Figure 4-1). Prepare a temporary wet mount from the culture provided. Note the chlorophyll contained in numerous **chloroplasts**. These lie just inside the plasma membrane. Does *Euglena* have a cell wall? The highly refractile bodies are plastids that function in carbohydrate storage. The nucleus is within the clear area near the center of the cell. Look for the orange-red **eyespot** near the anterior end. Can you see the flagellum? In active locomotion it may be coiled back around the body. Note that *Euglena* can move both by swimming and by creeping over a surface. Can it change the shape of its body? How? Would you call *Euglena* a plant or an animal?

II. CLASS SARCODINA

Members of this class move and capture food by means of **pseudopodia** (false + feet), which are extensions of the cytoplasm. All sarcodines are phagotrophic, that is, they feed by engulfing food into **food vacuoles**. Their nutrition is thus animallike. What metazoan cells resemble sarcodines? Sarcodines are closely related to certain flagellates. Some species alternate between a nonflagellated amoeboid form and a flagellated form.

A. **Amoeba**

This is a large sarcodine. Examine a preparation of living *Amoeba,* find an active specimen, and observe it for several minutes to determine its rate and direction of movement. It feeds on other protozoans by engulfing them with its pseudopodia. Examine your *Amoeba* for food vacuoles. Using high power, find the clear **ectoplasm** and the granular **endoplasm**. The endoplasm is composed of an external layer of gelated protoplasm, the plasmagel, and an internal layer of flowing protoplasm, the plasmasol. Find the transparent **contractile vacuole** and note its activity. This organelle removes excess water from the cell. What happens to the internal structures as the amoeba moves about?

In the space provided make a drawing, about 3 inches long, of an *Amoeba*. Label the structures you have seen. Indicate with arrows the direction of movement of the organism and the direction of flow of the plasmasol.

Amoeba, **movement**

Examine the demonstration slide of stained amoebae to see the nucleus. Add the nucleus to your drawing.

B. **Other Sarcodines**

Many sarcodines secrete hard exoskeletons or endoskeletons. Examine the demonstrations of *Difflugia* and *Arcella* (Figure 4-1), which are examples of sarcodines that have shells. The exoskeletons of numerous **Foraminifera** will also be on demonstration. Foraminiferans live in the oceans; when they die their shells settle to the bottom and may form sediments thousands of feet thick. The white cliffs of Dover are composed largely of the exoskeletons of these organisms.

III. CLASS CILIATA

These organisms move by means of **cilia**, which are short and very numerous cytoplasmic organelles that beat in a coordinated manner. Consult your textbook for a discussion of the mode of action of cilia. Morphologically, the ciliates are among the most highly differentiated protozoans. They have two types of nuclei, a **macronucleus**, which regulates metabolism and divides by amitosis, and a **micronucleus**, which controls the heredity of the organism and divides by mitosis or meiosis. Only the micronucleus participates in genetic recombination.

A. **Paramecium**

Paramecium (Figure 4-2) will be studied as an example of a ciliate. Note the rapid and coordinated swimming motion. How does *Paramecium* avoid obstacles? In the space provided, indicate the progression of a single individual and its reaction to an object in its path.

Paramecium, **swimming behavior**

Add a small drop of stained yeast cells to your wet mount. Note how the paramecia engulf them and form food vacuoles. What happens to the color of the food vacuoles? Why? A drop of methocel may be added to the preparation to slow down the rapid swimming motion before the cover slip is added.

Find the **oral groove** and **cytostome** (cell mouth). Function? Find the two contractile vacuoles. Do they contract synchronously or alternately?

Trichocysts are organelles of unknown function embedded in the **pellicle** (specialized plasma membrane) of *Paramecium*. When stimulated, they discharge to form twisted filaments. Examine the demonstration of discharged trichocysts.

Examine the stained demonstration slide of *Paramecium* showing the **micronucleus,** the **macronucleus,** and individuals undergoing **fission.** How do *Amoeba, Euglena,* and *Peranema* reproduce?

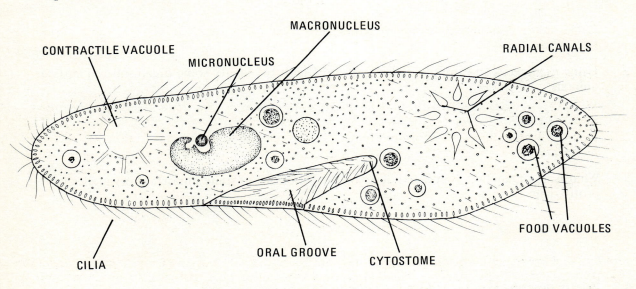

FIGURE 4-2 *PARAMECIUM*

B. **Other Ciliates**

Examine the demonstrations of other ciliates that may be available. *Vorticella* is a stalked ciliate. Note that the stalk may contract upon stimulation. The cilia near the cytostome have fused in membranelike rows. *Euplotes* has groups of cilia fused to form leglike bristles with which it walks over a substrate.

17

C. Conjugation in Ciliates

In addition to reproduction by binary fission, ciliates exhibit a unique process of sexual recombination called **conjugation**. Examine the demonstration slide of *Paramecium* conjugation.

In conjugation two cells of different mating types pair. The macronucleus in each individual degenerates, and the micronucleus undergoes meiosis. Three of the haploid nuclei so formed degenerate; the fourth undergoes a mitotic division. A cytoplasmic bridge forms between the conjugants, and one haploid gamete nucleus from each migrates through the bridge to the other conjugant. The exchanged gamete nuclei next fuse with the stationary gamete nuclei to form a new diploid micronucleus in each conjugant. At this time the two individuals break apart. The new micronucleus in each exconjugant divides by mitosis. One of the daughter nuclei develops into a new macronucleus; the other remains as a micronucleus.

IV. SPOROZOANS

These are protozoans that completely lack locomotor organelles. All are parasites. A well-known example is *Plasmodium vivax*, the organism that causes malaria. Examine the demonstration slides that show stages of malaria.

QUESTIONS

1. Compare the means of locomotion in the major groups of protozoans.

2. Using *Euglena* as an example, tell why it is difficult to classify some organisms as plants or animals.

3. What is an organelle? Name and describe the functions of some important organelles in protozoans.

4. Briefly describe asexual and sexual reproduction in *Paramecium*.

5. INTRODUCTION TO METAZOA: PHYLUM PORIFERA

I. METAZOA *multicellular organisms*

The multicellular animals, like other large and diverse groups of organisms, are divided into a group of phyla (singular: phylum) reflecting basic similarities of organization. Each phylum is interpreted as representing a unique structural plan. There are 20 to 30 such phyla of multicellular or **metazoan** animals (the variation representing lack of unanimity among zoologists in this matter). Of these we shall study nine, chosen both because of their importance and as examples of different patterns of animal organization.

Zoologists are agreed concerning the evolutionary relations among these groups, which are portrayed in Figure 5-1. This evolutionary pattern forms a background for our survey of animal groups. However, it should be emphasized that each group represents a functionally successful solution of the problem of metazoan organization, not a benighted and frustrated failure to achieve arthropod or chordate status.

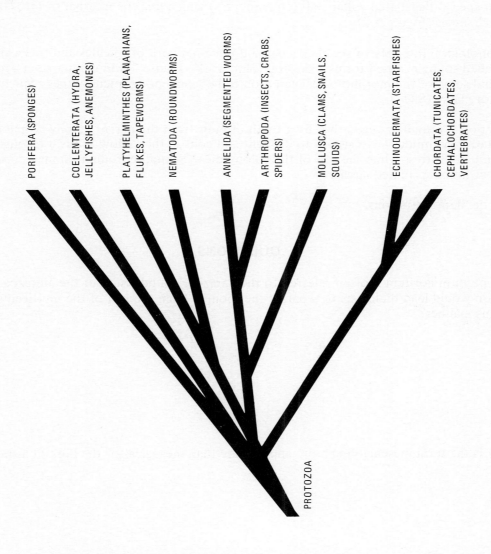

FIGURE 5-1 PHYLOGENETIC TREE OF ANIMAL PHYLA

II. SPONGES

Porifera (porus = pore, fera = bearing).

Approximately 5,000 species are known.

The first phylum of multicellular animals that will be studied is the Porifera. It consists of the sponges, which are among the most primitive of metazoan animals.

All sponges are aquatic, and most are marine. They are usually colonial and vary in size from a few millimetres to 1 or 2 metres in diameter. Their coloration is also highly variable from species to species, ranging from drab brown and gray to brilliant orange and red, although freshwater species are green or brown. These sessile animals grow in a variety of forms so that symmetry is either lacking or radial.

Sponges feed by filtering microscopic plant and animal matter from the surrounding water. Currents of water enter the animals through small pores (ostia) in the body surface because of the activity of special flagellated collar cells (choanocytes) which line the internal surfaces of the body cavity or channels (Figure 5-2).

The morphological integrity of sponges is based upon a network of **spicules** and (or) a structural protein called **spongin** which forms a skeletal framework. Figure 5-3 illustrates such a network of spicules and some of the morphological variation to be seen among spicule types. Spicules are limy ($CaCO_3$) or glassy (SiO_2).

Reproduction is primarily vegetative, by growth of buds from the parent colony or **gemmules**. The vegetative gemmules are resistant to drought and cold so they allow survival through periods of environmental stress while the rest of the colony dies. Sexual reproduction may occur to produce free-swimming ciliated larvae.

Examine the demonstrations.

QUESTIONS

1. The phylum Porifera is often referred to the Parazoa, a subdivision of the Metazoa. What reason would lead biologists to separate the sponges from the rest of the multicellular animals in this manner?

2. Why is the term **mesenchyme** more appropriate than mesoglea for the bulk of a sponge body?

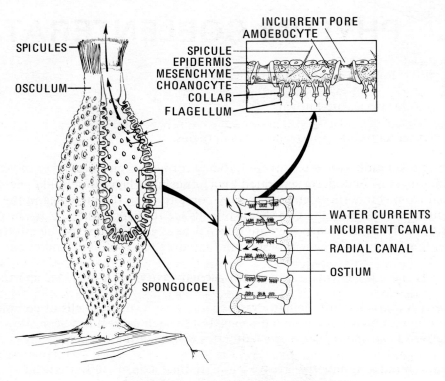

FIGURE 5-2   DIAGRAM OF SIMPLE SPONGE

FIGURE 5-3a   Skeletal remains of a freshwater sponge. The spherical mass of spicules covers a gemmule. Photograph by Dr. Louise Rollins.

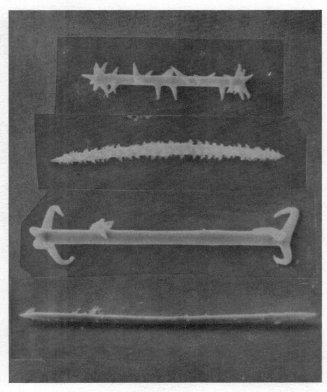

FIGURE 5-3b   Morphological variation of some freshwater sponge spicules. The lengths of these types of spicules vary between 50 μm and 200 μm. Photograph by Dr. Louise Rollins.

6. PHYLUM COELENTERATA

Coelenterata (koilos = hollow, enteron = gut).

Approximately 10,000 species are known.

The phylum Coelenterata has the following characteristics.

1. The body is **diploblastic** (two-layered), being formed of an outer layer of ectoderm and an inner layer of endoderm separated by a noncellular layer of jellylike mesoglea. Cells derived from ectoderm or endoderm may come to lie in the mesoglea in the large coelenterates, but only the two sheets of cells mentioned contribute to the development of adult form. Hence they are referred to as **germ layers**. Some are **triploblastic**, with a third germ layer called mesoderm that gives rise to muscle and connective tissues.

2. The body has a single cavity, the **gastrovascular cavity**, with only one aperture, the mouth.

3. All possess tentacles with **nematocysts**, characteristic stinging cells of complex structure. Some other animals possess nematocysts, but in such cases it has been shown that they are derived from the tissues of coelenterates on which they prey.

4. **Radial** or **biradial symmetry** around a longitudinal axis; no definite head.

The phylum Coelenterata contains three classes.

Class Hydrozoa. Freshwater types such as *Hydra* and marine types such as *Obelia* and *Gonionemus*. Marine forms typically show an alternation between a sexually reproducing **medusa** stage and an asexually reproducing **polyp** stage.

Class Scyphozoa. Marine jellyfishes. The medusa is enlarged and gelatinous. The polyp is absent or appears only as a larval form.

Class Anthozoa. Sea anemones and corals. These animals have no medusoid form.

In our study of coelenterates we will emphasize three members of the class Hydrozoa: *Obelia, Gonionemus,* and *Hydra*.

I. OBELIA

The structure and life history of *Obelia* show two basic characteristics of the coelenterates. These are:

1. The tendency for the same organism to have more than one form (polymorphism).

2. The occurrence of a regular sequence of polypoid (hydralike) and medusoid (jellyfish-like) forms, or morphs, in the life cycle.

Obelia has three body forms (hydranth, gonangium, and medusa). Other coelenterates may have as many as seven or eight. The interpretation of the occurrence of different forms in the life cycle will be discussed later.

A portion of a colony of *Obelia* is shown in Figure 6-1.

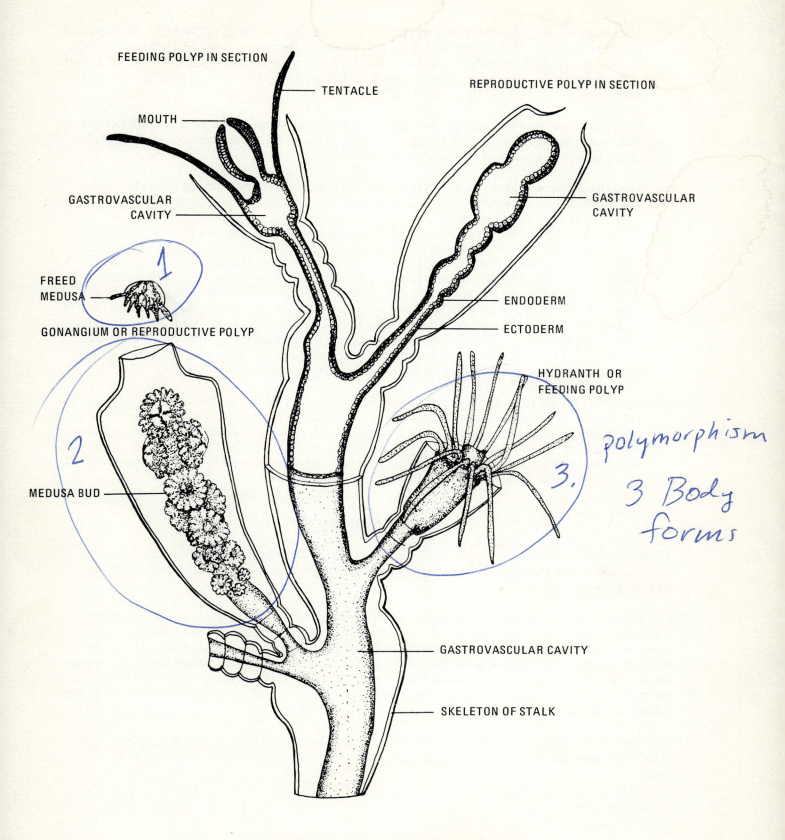

FIGURE 6-1 *OBELIA*, HYDROID COLONY

A. Note the presence of a stalk from which new individuals, called **polyps,** bud off alternately to the right and left. Find the two types of polyps which are present: one with tentacles, the hydroid polyp, or **hydranth**; one without tentacles but with **medusa** buds surrounding a central stalk, the reproductive polyp, or **gonangium**. What activities does each type of polyp carry out?

Each polyp has **ectoderm**, a thin **mesoglea**, a thicker **endoderm**, and a **gastrovascular cavity**. Note that this cavity joins with that in the stalk. A common and continuous gastrovascular cavity is therefore present throughout the colony.

Find the chitinous exoskeleton which surrounds the colony. The individual polyps can retract within this exoskeleton, giving it a protective as well as a supportive function.

B. The life history includes both the colonial hydroid stage, which you see here, and a solitary medusoid stage. The **medusae**, which are small, escape and mature away from the colony. Some are functionally male, other female. Sperm and ova are discharged into the water, where fertilization and development to the early **hydroid** stage occur. The early hydranth puts out a bit of stalk, which forms a second hydranth. This hydranth puts out a stalk on the other side, to form a third hydranth, and so on until a large number of individuals are present in the colony. When a colony is mature and well nourished, secondary buds come from the base of the stalk of many hydranths and develop into reproductive polyps called **gonangia**. A number of tiny jellyfish (medusae) bud off each gonangium. See the demonstration of mature medusae and of a young medusa emerging from a gonangium.

Note that sexual reproduction is the exclusive prerogative of the medusa individuals while the production of the polypoid members of the colony from the original hydroid stage is the result of the asexual process of budding. Hence the life cycle is basically an alternation between an asexual polyp phase and a sexually reproducing medusa. In various coelenterates, one or the other of these phases may be conspicuous and the other minute or absent altogether. *Gonionemus* and *Hydra* are examples of such modifications.

II. GONIONEMUS

The hydrozoan "jellyfish" *Gonionemus* (which has a small hydroid stage) will illustrate the plan of organization of the medusa. The same germ layers are present as in *Hydra* and *Obelia*, but the mesoglea is much thicker.

The gastrovascular cavity is limited to the **mouth**, surrounded by the **oral lobes**; a small **stomach**; the **radial** canals; and the **circular** canal. The **gonads** hang down in folds beneath the radial gastrovascular canals. Numerous tentacles which act in the capture of food animals radiate from the edges of the bell. The **bell**, or **umbrella**, of the jellyfish has an upper **ectodermal** layer, the **exumbrella**, and a lower layer, also of **ectoderm**, the **subumbrella**. Between these is the thickened **mesoglea**. A **velum**, or diaphragm, is a shelf of tissue which extends inward from the region of the circular canal. See Figure 6-2.

The medusa moves by the principle of jet propulsion. The contractions of the bell, produced by ectodermal muscles, force water out past the velum with some force, moving the bell through the water.

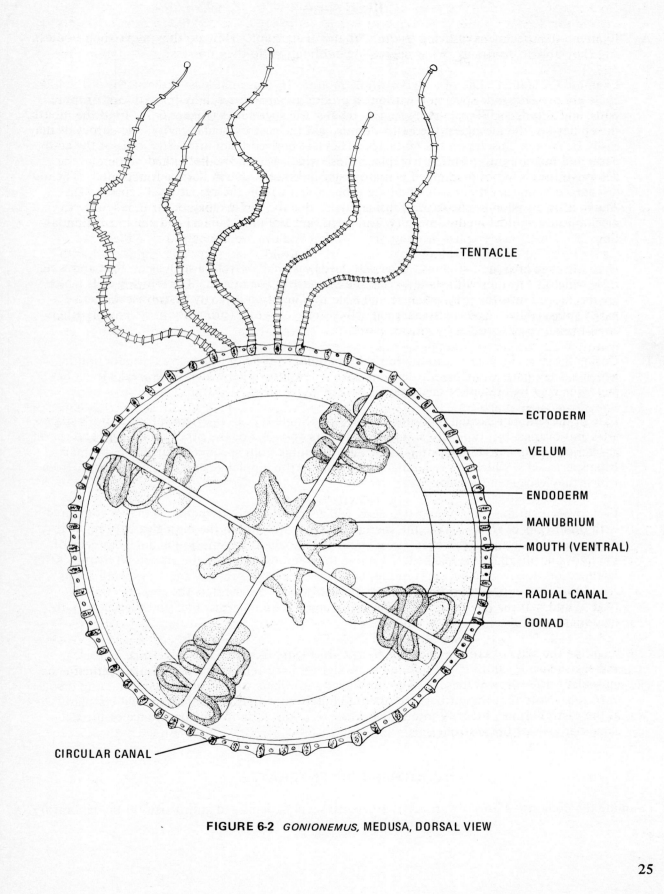

FIGURE 6-2 *GONIONEMUS,* MEDUSA, DORSAL VIEW

III. HYDRA

A. Examine the specimens of living *Hydra*. What is their color? How do they move their bodies? Are they able to contract? What degree of coordination do they possess?

Examine the stained slide of a *Hydra* whole mount. Hydras in at least three stages of the life cycle are present: one specimen without reproductive structures, a hydra with one or more **buds,** and a hydra with male reproductive organs, the **spermaries** (Figure 6-3). Find the mouth, the epidermis, the mesoglea, the gastrodermis, and the gastrovascular cavity. Nematocysts may easily be seen in clusters on the tentacles. They are derived from interstitial cells of the ectoderm and migrate into position to replace those which have been discharged. Examine the demonstration slide of discharged nematocysts, and sketch one in the space provided. The undischarged nematocyst consists of a long tube, coiled within the nematocyst capsule. Discharge is by eversion or the turning inside out of this tube. This may occur in response to mechanical stimuli alone but more typically requires both mechanical and chemical stimulation.

Hydra lives in lakes and streams. It usually feeds on small crustacea such as cyclops and water fleas which it captures with its tentacles and crams into the mouth. The stinging cells which are discharged into the prey paralyze and hold it. **Digestion** is partly **gastrovascular** and partly **intracellular**. Some gastrodermal cells secrete enzymes (primarily proteinases); others engulf small food particles by phagocytosis.

B. During most of the year, reproduction is by **budding,** a type of vegetative reproduction in which a small part of the parent develops into a new individual which then breaks free. Several buds may be present at one time.

Late in the summer, sexual reproduction occurs. Some species of *Hydra* produce both spermaries and ovaries, but typically an individual bears only one or the other. Both eggs and sperm are derived from ectodermal interstitial cells. Examine your specimen showing a spermary. Ripe sperm cells, which are very small, are found in the nipplelike end of the spermary, from which they escape in swarms.

Each ovary contains only one or two eggs, which are relatively tremendous in size compared with other cells of *Hydra*. Like the sperm cells, they develop in the mesoglea between ectoderm and endoderm. There they move about by amoeboid locomotion and devour cells of the hydra until full size is reached. Examine the demonstration of the amoeboid ovum. After fertilization, the zygote becomes round, begins the cleavage process, and a cyst wall forms around it. The encysted embryo is a resistant stage that overwinters the species. Unlike typical hydrozoans, *Hydra* lives in fresh water and has no medusa. Can you suggest why the medusa stage is absent?

C. Examine the slide of cross sections of *Hydra*. In Figure 6-3 label the epidermis, mesoglea, and gastrodermis. Both epidermis and gastrodermis have cells which are termed **epitheliomuscular**. The bases of these cells are produced into fibers which form the contracting tissue in *Hydra*. Note that these fibers are oriented longitudinally in the epidermis and circularly in the gastrodermis. Identify interstitial cells. Note the variation in appearance of the tall glandular cells of the gastrodermis.

IV. OTHER COELENTERATES

Examine the demonstrations of various hydrozoans, scyphozoans, and anthozoans in the laboratory.

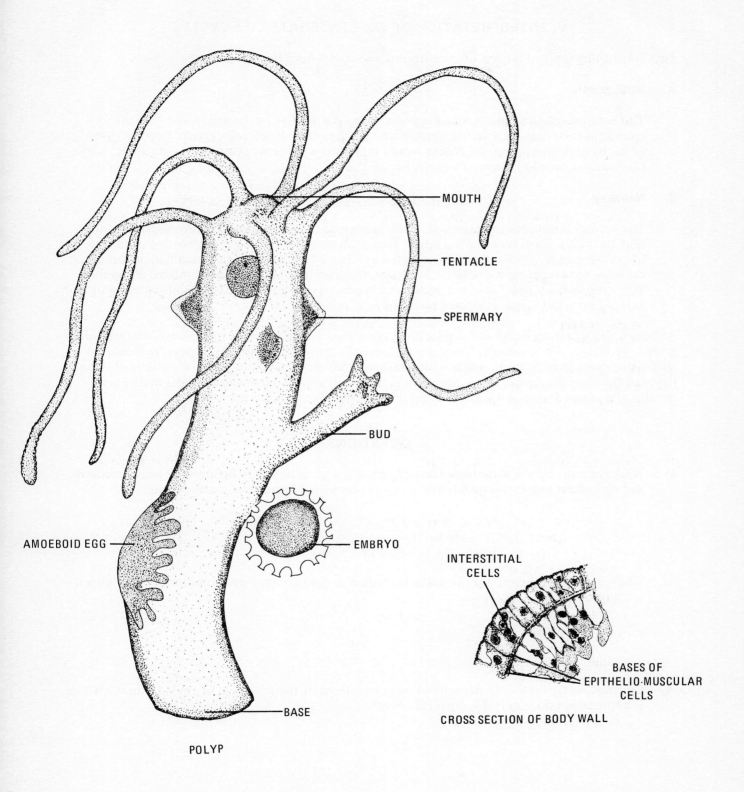

FIGURE 6-3 *HYDRA*

V. INTERPRETATION OF COELENTERATE LIFE CYCLES

Two contrasting interpretations of coelenterate life cycles have been proposed:

A. Metagenesis

This interpretation suggests that the polyp form is primitive in the group and that the medusoid form was secondarily derived and the function of sexual reproduction relegated to it. By this interpretation, *Hydra* would represent a primitive form of the life cycle, while *Gonionemus* would represent a derivative of this.

B. Neoteny

The second interpretation argues that the medusa is the adult form of these organisms and that the polyp form represents a larva. Hence *Gonionemus* would represent the primitive life cycle, and *Hydra* would be interpreted as a persistent larval form which had acquired the capacity for sexual reproduction. In *Hydra,* the primitive adult form is omitted altogether. This evolutionary process based on the development of a persistent larva and omission of antecedent adult stages is termed **neoteny**.

Although this question is still debatable, most zoologists are agreed that the second interpretation is more probable. For example, *Gonionemus* has only one rather primitive type of nematocyst; *Hydra* possesses four different kinds with a rather highly developed and specific pattern of stimuli required for their discharge. Also, the mere fact that *Hydra* is a freshwater animal makes one question its status as a stem form for so predominantly marine a group.

QUESTIONS

1. What is meant by the term **diploblastic**? What is a germ layer? Why are the terms **ectoderm** and **endoderm** employed differently from the terms **epidermis** and **gastrodermis**?

2. What is polymorphism? Of the animals studied in detail, which are polymorphic and which are monomorphic?

3. Contrast metagenesis and neoteny as interpretations of the life cycle of coelenterates. Which interpretation do you feel is correct? Why?

4. A muscle cell can only contract; hence the virtual universality of antagonism between muscles across joints. Explain how the epitheliomuscular layers in *Hydra* are antagonistic in action.

7. PHYLUM PLATYHELMINTHES

Platyhelminthes (platys = flat, helmins = worm).

Approximately 15,000 species are known.

The phylum Platyhelminthes includes the flatworms. Some are free-living but many are parasitic. The phylum has the following characteristics.

1. The presence of three germ layers.

    a. Ectoderm, gives rise to the epidermis.
    b. Endoderm, gives rise to the epithelium of the digestive tract.
    c. Mesoderm, gives rise to organs that solidly fill the space between the ectoderm and the endoderm. There is no body cavity.

2. The presence of organ systems.

    a. Digestive system, with a mouth but no anus.
    b. Excretory system, of **flame cells** and ducts.
    c. Nervous system, of "brain" and solid cords.
    d. Muscular system.
    e. Reproductive system.

3. **A flattened body,** with a definite **anterior end** and **bilateral symmetry** (right and left sides essentially mirror images).

The phylum Platyhelminthes is usually divided into three classes.

Class Turbellaria. Free-living flatworms.

Class Trematoda. Parasitic flukes.

Class Cestoda. Parasitic tapeworms.

I. CLASS TURBELLARIA

A flatworm commonly called planaria will be studied. This worm, which reaches one-half to three-quarters of an inch in length, is found in ponds or sluggish streams under sticks, stones, and bottom debris. It eats dead or decaying animal and vegetable matter (Figure 7-1).

A. Examine the living planaria for size, shape, color, and activity. Locomotion is produced through the coordinated action of the millions of cilia covering the body, whereas the direction of movement is under muscular control.

    Try turning the animal on its back. How does it right itself? *spiral like twisting* What is its reaction to bright light? To darkness?

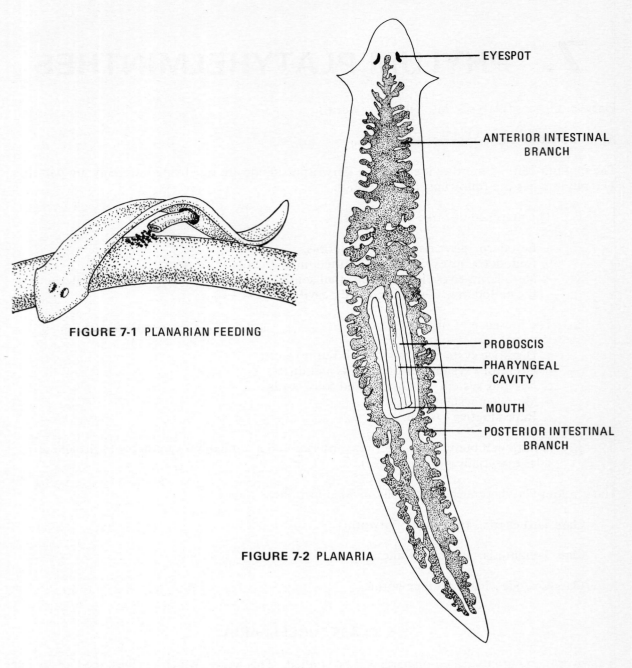

FIGURE 7-1 PLANARIAN FEEDING

FIGURE 7-2 PLANARIA

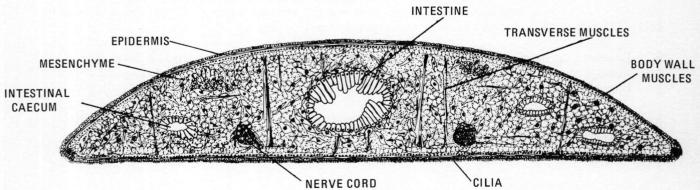

FIGURE 7-3 PLANARIA, CROSS SECTION

The planarian senses food through **chemical sense receptors** located primarily in the regions of the head and proboscis. Concentrations of these and other receptors occur in the tip of the head, in the lateral earlike lappets, and under the light-sensitive **eyespots**. Response is coordinated through nervous action controlled through the "brain" between the eyespots and at least one pair of **nerve cords** running the length of the body. This system is difficult to see in the living animal and is not stained in the permanent slides. Special staining methods are necessary to show it.

B. Examine the permanent slide of planaria (Figure 7-2). This animal was fed meat ground with lampblack before it was killed, stained, and mounted on the slide. The lampblack outlines most of the **digestive tract,** which consists of a **mouth**, a **pharynx**, and an **intestine** of three branches: one anterior and two posterior. The branches are subdivided into the intestinal ceca (singular: cecum).

The prominent **proboscis** can be protruded through the mouth, which is located on the mid-ventral surface of the body, *not* in the head (Figure 7-1). The pharynx, running through the proboscis, connects the mouth with the intestine. Except when planaria is feeding, the proboscis is withdrawn into a cavity in the lower surface of the body. Locate the eyespots near the anterior end. Most other structures are not distinguishable in the whole mount.

C. The cross sections of planaria were taken through three regions: the anterior region, with a single intestinal branch and some of the intestinal ceca; the middle region, through the pharynx; and the posterior region, with the two main intestinal branches and ceca. The first section is outlined in Figure 7-3. Note the **epidermis** (ciliated **ectoderm**, one cell in thickness), the layer of circular and longitudinal muscles (mesoderm) just beneath, the two **intestinal branches** and **ceca** (of a single layer of columnar **endodermal** cells), and the loose, fluid-filled **mesenchyme** (mesoderm) in between. The nerve cords, reproductive tissue and related organs, excretory structures, and **transverse muscles** are present in the mesenchyme. Except for the transverse muscles, you need not identify these structures.

Respiration in planaria is accomplished directly through the body surface. There are no special respiratory organs. **Excretion** occurs through the action of special cells, called **flame cells,** which occur in the flatworms and a few other groups, including the rotifers. The cell contains a bladder, which communicates with a duct leading eventually to the outside. In the bladder a tuft of cilia ("flame") beats continually, driving the fluid which comes into the cell from the tissue, out the duct. The hundreds of flame cells in planaria (there are fewer in smaller animals) act as a filtration plant, removing wastes and excess water from the body.

Reproduction is both asexual and sexual. In **asexual reproduction**, transverse division of the body occurs. A new head begins to form about the middle of the body (just behind the pharynx), and eventually a constriction occurs in front of the new head. In simpler flatworms, chains of individuals occasionally are produced. If available, dividing *Microstomum* will be demonstrated. The remarkable ability of the flatworms to regenerate following injury is correlated with their ability to divide asexually. One simple experiment will be performed in the laboratory. The instructor will cut two or three planaria in half, through the pharynx, and keep the parts together in a dish long enough for regeneration to occur. The anterior halves will each form a new pharynx and a new posterior region. The posterior halves will each form a new head and a pharyngeal apparatus more slowly. Sooner or later, two normal planaria will be present where one was before. Much smaller pieces will regenerate all missing parts, but the process takes longer.

Sexual reproduction involves the production of gametes. Usually both eggs and sperm are produced in a single individual, and sperm cells are then exchanged at mating. The flatworms, therefore, are **bisexual** or **hermaphroditic**. Because the reproductive system and its operation is not readily studied in planaria, a small fluke *(Opisthorchis)* will be used instead.

II. CLASS TREMATODA

Clonorchis (Figure 7-4) is the human liver fluke of the far east. The genus *Opisthorchis* includes a similar fluke found in the muskrat (Figure 7-5).

A. General Anatomy

Consult Figures 7-4 and 7-5 and identify the labeled structures from the specimen on your slide.

B. Reproduction

The adult *Opisthorchis* lives in the bile duct of a muskrat, where food is abundant, and needs little more than a reproductive system to survive. That system is very well developed. Its operation, in brief, is as follows: **Sperm cells**, produced in the **testes**, make their way through the sperm ducts (usually invisible) to the **seminal vesicle**, where they are stored. When two worms copulate, sperm cells are exchanged and are stored in the **seminal receptacles**. Mating usually occurs early, before egg production, and the sperm cells are carried to the receptacle through the empty uterus. **Ova** are formed in the **ovary** and are discharged into the base of the **oviduct** where fertilization occurs. Yolk is then added and finally a **shell** is deposited on the egg as it moves into the part of the oviduct called the **uterus**. Here the eggs accumulate until a large number are present.

The eggs are discharged continually from the **genital pore** and are eventually carried out of the body of the host, usually in the intestinal wastes.

C. Life Cycle

Before the adult stage of a fluke again occurs in the final host, several larval stages occur elsewhere. As a general rule, a fluke egg, when deposited in water, hatches into a tiny ciliated larva (miracidium). To survive, the miracidium must penetrate the body of a particular snail, where it grows into a comparatively huge second larval stage (sporocyst) with great reproductive capacity. Often additional larval stages (rediae) occur in the snail tissues, but eventually large numbers of the final larval stage (cercariae) are produced. The cercaria bores out of the snail and swims by means of its tail (cercus = tail) for a short time. It may penetrate the body of the final host directly, or else encyst in or on vegetable or animal food habitually eaten by the final host. When eaten, the larva escapes from its cyst wall and develops to maturity after working its way to the bile duct. The life cycle involves many hazardous events for the parasite, for which compensation is made in reproductive capacity vastly increased over that of free-living forms.

Note that the life cycle involves reproduction by one or more of the larval stages. In some cases this occurs to an impressive extent, a million cercariae being liberated from a snail invaded by a single miracidium. This reproduction in the larval stages of the flukes is asexual. The general term for this process is **paedogenesis** (child reproduction).

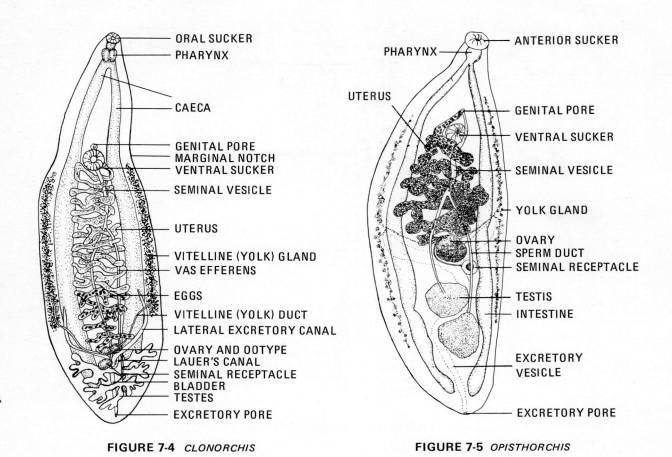

FIGURE 7-4 CLONORCHIS   FIGURE 7-5 OPISTHORCHIS

III. CLASS CESTODA

The tapeworm *Dipylidium,* which occurs commonly in dogs and cats and rarely in children, will be studied.

A. General Anatomy

The whole tapeworm is termed a **strobila**. It may be studied using the slide provided. The strobila consists of a **scolex** (Figure 7-6) bearing four suckers and several hooks, followed by a chain of repeated sections termed **proglottids** (Figure 7-7).

1. SCOLEX

Observe the suckers, spines, and hooks on your specimen. The adult tapeworm is found in the small intestine. The scolex serves for attachment to the intestinal lining. It does not function in the nutrition of the tapeworm.

2. PROGLOTTIDS

Proglottids are budded off in a region just behind the scolex and undergo a process of maturation. Immature proglottids are wider than long. As maturation proceeds, the reproductive organs become apparent. The terminal proglottids are termed **gravid** and are essentially sacs of fertilized eggs. Gravid proglottids are broken off the terminal portion

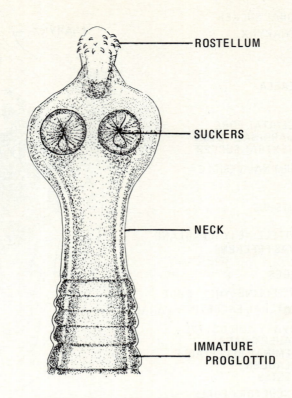

FIGURE 7-6 TAPEWORM, SCOLEX

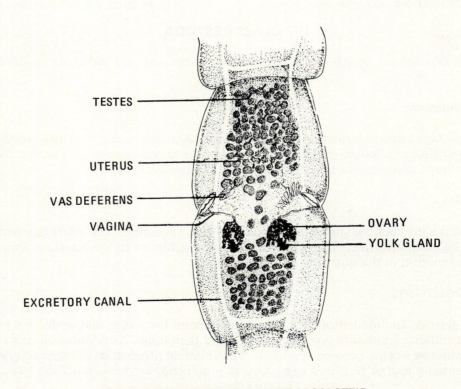

FIGURE 7-7 TAPEWORM, MATURE PROGLOTTID

of the strobila. Sketch in outline form an immature and a gravid proglottid. With the aid of Figure 7-7 identify the structures in the mature proglottids.

Tapeworms absorb food from the intestine of the host directly across their body wall and consequently show no sign of a digestive tract.

B. Life Cycle

The life cycles of tapeworms vary considerably, but that of *Dipylidium* is reasonably representative. When mature proglottids are shed from the strobila, they are passed to the exterior in the feces of the dog or cat. There they may be eaten by larval stages of one of several fleas, usually the cat flea. Many of the flea larvae die as a result of the development of the tapeworm larva, but some survive and develop to adult fleas. When these are eaten by their host, the tapeworm larva develops to an adult organism in the small intestine. Note that here, as in the flukes, there is an elaborate reproductive system capable of producing immense quantities of eggs. However, it is worth noting that the reproductive system of flukes and tapeworms is not qualitatively different from that found in free-living flatworms; rather it is simply more prominent and, in the tapeworms, repeated serially.

QUESTIONS

1. What is the difference between the mesoglea of the coelenterates and the mesenchyme of the flatworms?

2. Trace the pathway followed by sperm in the process of fertilization in *Opisthorchis*. What must be the sequence of events in the production of the shelled eggs containing cellular yolk?

3. It can be argued that the sporocyst and redia stages in the life cycle of a fluke are comparable to the polyp stage in *Obelia*. Explain the similarities.

4. What is a parasite? What is a host?

5. Compare the concepts of neoteny and paedogenesis.

8. PHYLUM NEMATODA

Nematoda (nema = thread, eidos = form).

Approximately 15,000 species are known.

The phylum Nematoda includes the **roundworms**. The nematodes are highly successful. They occur abundantly in virtually every kind of habitat. Many are parasites, and most kinds of animals and plants harbor nematodes. If all the species of nematodes were known, they might outnumber even the arthropods, a phylum for which about 1 million species have been described. A few nematodes are large but most are small, even microscopic.

The phylum Nematoda has the following characteristics.

1. The presence of three germ layers.

    a. Ectoderm, the dermal syncytium that secretes a **cuticle** covering the body.
    b. Endoderm, the epithelium composing the whole digestive system.
    c. Mesoderm, in muscles, excretory, and reproductive organs.

2. The presence of organ systems.

    a. Digestive system, a straight tube with mouth and anus.
    b. Excretory system, of tubules and two main lateral canals.
    c. Nervous system, a "brain" encircling the pharynx and two main solid nerve cords, one dorsal and one ventral.
    d. Muscular system, longitudinal muscles only.
    e. Reproductive system, sexes separate.

3. A body round in cross section with definite anterior end and bilateral symmetry.

4. A fluid-filled body cavity termed a **pseudocoel** between the muscles (mesoderm) and the gut (endoderm).

I. ASCARIS

This large nematode is a parasite in many mammals, including humans. Laboratory specimens are typically from pigs.

A. **General Anatomy**

Females are larger than males and males have a hooked posterior end. Dissected specimens might be on demonstration or you might dissect one or both sexes. If dissecting, carefully cut the worm along its entire length with a sharp edge such as a safety razor blade, pin the worm out in a dissecting pan, and add enough cold water to keep the specimen moist.

Compare your specimen with the demonstration model of a typical nematode.

1. Female

    In the female, locate: **digestive tract, ovary, oviduct, uterus,** and **vagina** (see Figure 8-1).

2. Male

    In the male, locate: **digestive tract, testis, seminal vesicle,** and **copulatory spicules.**

B. Cross Section

Examine the sections of *Ascaris*. There are three pairs of sections on the slide: one through the anterior region of the worm (Figure 8-2), and two pairs through the reproductive organs of the two sexes. At first confine your attention to the simplest section. Identify the cuticle, apparently of several distinct layers, and the **dermal syncytium** beneath (dermal = skin, syncytium = a tissue with many nuclei and few cell boundaries). The dermal syncytium extends into the body cavity on each side, where it contains the lateral **excretory canals**. These canals open to the outside in the anterior region of the body.

The muscle cells in nematodes are entirely unique. Each cell has a large expanded portion that contains the nucleus and extends well into the body cavity and a basal portion that bears columns of myofibrils. Cut in cross section, the latter look like striated bars. Extending from the expanded nuclear portion of the cell to the dorsal or ventral nerve cord are processes of the muscle cells, filamentous outgrowths which effect the junction with the nervous system.

The movement of nematodes is a characteristic threshing produced by alternate contraction of the dorsal and ventral muscles. Note that only longitudinal muscle is present.

Interrupting the muscle layer dorsally and ventrally (you cannot distinguish which is which) are two ill-defined **nerve cords** with branches at irregular intervals. Other nerves also occur.

Find the intestine in the center of the pseudocoel. The intestine is composed of a single layer of epithelial cells. Internally, the cells possess a **brush border**, which looks as though it is composed of cilia. However, the processes are nonmotile and matted together. They function to increase the absorptive surface area of the intestine.

Color the various organs pictured in Figure 8-2 according to the germ layer from which they originate. Tissues derived from ectoderm should be colored blue, those from mesoderm red, and those from endoderm yellow.

Identify the regions of the male and female reproductive system in the appropriate cross sections on your slide.

1. FEMALE

    a. The **uterus** can be identified as a thin-walled tube filled with eggs.
    b. The **ovary** is composed of a central cord around which developing eggs are arranged like spokes of a wheel. The developing eggs appear as greatly elongated epithelial cells.
    c. The **oviduct** wall looks like that of the ovary, but the central cord is not present.
    d. Reexamine the dissected specimen and account for the multiple sections of uterus, ovary, and oviduct that appear on your slide.

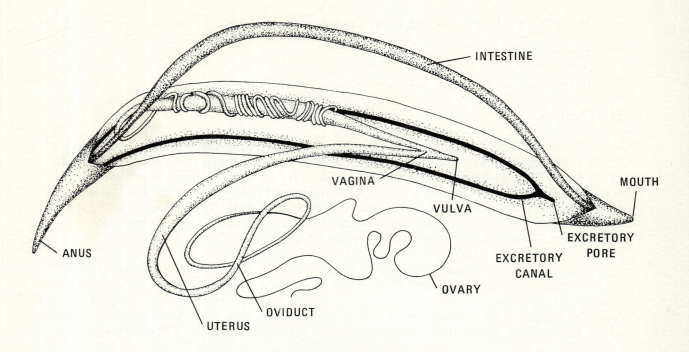

FIGURE 8-1 FEMALE *ASCARIS*, GENERAL ANATOMY

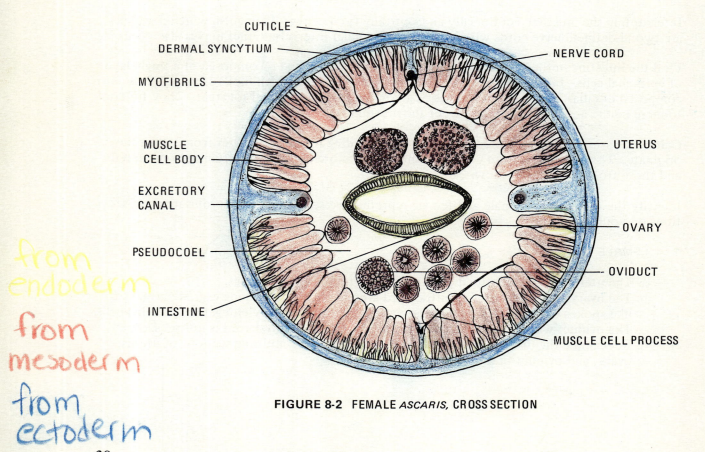

FIGURE 8-2 FEMALE *ASCARIS*, CROSS SECTION

from endoderm
from mesoderm
from ectoderm

2. MALE

   a. Many sections of the **testis** containing amoeboid sperm in various stages of development will be found. Depending on the level of the section, either the large, thin-walled **seminal vesicle** or the muscular ejaculatory duct can be found. Both are filled with sperm.

II. VINEGAR EEL

This tiny nematode will be studied alive, particularly for activity and details of reproduction. It is found in vinegar, where it lives in concentrations of acetic acid fatal to most organisms. It feeds on the bacteria responsible for the formation of the acetic acid. It is also found in decaying apples.

The "eels" may be studied with small difficulty if they are embedded in debris on the slide. Otherwise, they are too active. It was from this rapid and continuous movement that the organism was given the name *Turbatrix* (turbo = to disturb). Observe a few worms under low power for a few moments until you are satisfied you understand their method of locomotion. Their movement can be brought to a standstill with heat, making study much easier. Oral directions will be given as to the simplest method of heat anaesthetization.

First examine a number of worms, for **body form, body cavity** (which may be more or less completely filled with fat droplets), digestive tract (**mouth, pharynx, intestine, anus**), and sex. The males have curved tails and prominent **spicules** near the tail region (Figure 8-3); the females, which are larger, lack these structures. Most of the females will be pregnant; they will contain larval worms ready to be born and embryos in various stages of development.

Select a female with a **few** embryos, and identify as many as possible of the structures labeled in Figure 8-4. Reproduction, which is sexual only, occurs as follows: As mating takes place, **sperm cells** from the male are deposited in the **seminal receptacle** of the female. Some of the sperm cells migrate through the **uterus** and the **oviduct** to the point where eggs are released from the ovary. After fertilization, the zygote is carried through the oviduct to the uterus, where it enlarges through the addition of yolk until full size is reached. Thereupon, cleavage follows. You will be able to find the **1-, 2-, 4-, 8-,** and **16-cell stages,** and several stages in the development of the larva. You may even be able to watch the division of the cells in one of the early cleavage stages. Given sufficient time, you could follow the complete development of a single individual.

III. OTHER NEMATODES AND ROTIFERS

Examine the demonstration slides of *Trichinella,* the organism that causes trichinosis, and of *Necator,* the hookworm. The nematodes are an extremely important group of parasites of both plants and animals. Life cycles vary in complexity from simple cycles, such as those of *Ascaris* and *Trichinella,* to cycles involving two hosts, periodic alternation of parasitic and free-living stages, and other elaborations.

Other phyla besides the Nematoda have a pseudocoel. An example is the phylum **Rotifera**. Examine the culture of rotifers. Rotifers have the following conspicuous features that you should recognize.

1. The **wheel organ.** This is a ciliated structure that creates the optical illusion of a turning wheel at the anterior end of the organism.

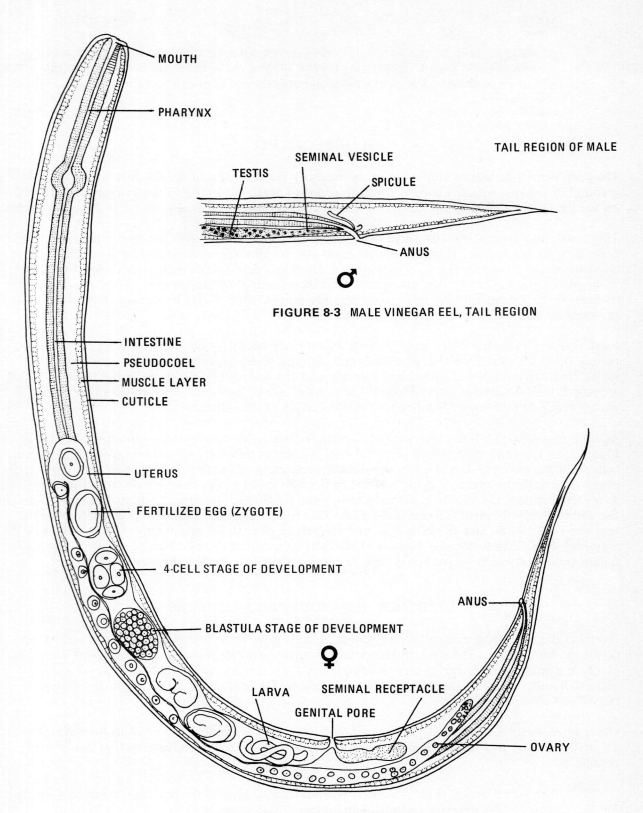

FIGURE 8-3 MALE VINEGAR EEL, TAIL REGION

FIGURE 8-4 FEMALE VINEGAR EEL

2. The specialized pharynx. This can be recognized by the rapidly moving, toothlike structures that serve as a grinding mechanism.

3. The tapering foot, which contains cement glands for attachment of the animal. Observe the feeding and locomotion of your specimen.

One of the interesting phenomena exhibited by the rotifers is that the number of cells composing an individual is constant for the species.

QUESTIONS

1. What is a pseudocoel?

2. Trace the pathway of sperm in the process of fertilization in *Ascaris*.

3. Assume that a female *Ascaris* produces 100,000 eggs per day and lives for one year. Calculate the probability on a population basis that a given egg will complete the life cycle and become a breeding adult. Hint: Population size remains relatively constant.

4. **Ovo-** refers to egg, **vivi-** means alive or active, **-parous** refers to parturition or birth. *Ascaris* is termed oviparous; *Turbatrix* is ovoviviparous; mammals are viviparous. Explain these terms and their application.

9. PHYLUM ANNELIDA

Annelida (annellus = little ring, -ida = plural suffix).

Approximately 8,500 species are known.

The phylum Annelida contains the **segmented worms,** including the familiar earthworm. The phylum has the following characteristics.

1. The presence of the three germs layers, ectoderm, mesoderm, and endoderm.

2. The presence of a **true coelom,** completely bounded by a mesodermally derived epithelium termed the **peritoneum.**

3. Well-developed organ systems as established in the nematodes, plus a circulatory system of **pulsatile vessels,** collecting and distributing vessels, and capillaries.

4. A segmented body, composed of an anterior prostomium (not a segment) followed by a series of segments. These segments are units that show **serial homology**; that is, they are basically alike in origin and structure. Segmentation in the annelids is termed **homonomous** because of the similarity of adult form and function of segments. In other groups of animals, segments may diverge considerably in their development.

    Primitively in the annelids, each segment as a unit bears a functional division of the following organ systems: excretory, nervous, muscular, reproductive, and circulatory. In addition, each segment bears a pair of coelomic pouches and a pair of appendages.

5. Paired appendages called parapodia. Parapodia bear bristlelike structures termed setae.

The phylum Annelida contains three common classes.

Class Polychaeta. Mostly marine worms. They have prominent parapodia and setae. They are considered primitive, the other classes having specializations for life on land and in fresh water.

Class Oligochaeta. Mostly terrestrial or freshwater forms. Parapodia have been lost and setae are small. This class includes the common earthworm.

Class Hirudinea. Leeches. Both parapodia and setae have been lost. Prominent anterior and posterior suckers are present.

I. NEREIS

The sandworm, *Nereis,* is a typical polychaete and shows basic annelid characteristics.

A. Head

The head of annelid worms is composed of the **prostomium** (not a true segment) and the **peristomium** (the first true segment). The head is well provided with sensory structures in *Nereis.*

Prostomium. Note the two prostomial tentacles and the pair of ventral palps. Four eyes lie dorsally on the posterior half of the prostomium.

Peristomium. Bears four pair of peristomial tentacular cirri, two pair dorsal in position and two pair ventral.

The mouth opens ventrally between the prostomium and peristomium. Sometimes in preservation the proboscis is everted as shown in Figure 9-1. Find such a specimen and observe the toothed chitinous jaws. As their presence suggests, *Nereis* is predatory.

B. **Parapodia**

Using a pair of sharp scissors or a safety razor blade, cut a single segment from the posterior one-third of your worm. It is well to make several preparations and compare them. Place them on a microscope slide with a little water and study them, first with the hand lens and then with the **low power** of the microscope. Identify the structures in Figure 9-2.

The parapodia have two main axes, one dorsal and the other ventral. In *Nereis,* each of these is supported by a stiffening rod and is further subdivided into lobes. Find the dorsal and ventral tentaclelike cirri.

In *Nereis* these appendages serve for locomotion and as respiratory organs.

C. **Other Polychaetes**

Although the head and the parapodia are basically comparable to the structures observed in *Nereis,* adaptive modifications have produced an astonishing variety of form. Examine the specimens of various polychaetes on display illustrating some of this variety.

II. EARTHWORM

A. **External Anatomy**

Distinguish between the dorsal and ventral surfaces. The ventral surface is somewhat flattened. Find the **mouth** just ventral to the **prostomium** and the anus at the posterior end. Segments 31 to 37 are swollen to form the **clitellum** (= pack saddle). This structure secretes a cocoon that slips off over the head of the worm after receiving eggs and sperm. Young worms develop in the cocoon.

Rub your fingers over the ventrolateral surfaces of the worm. Can you feel the **setae**? In what direction do the setae point? What is their function?

The openings of some of the reproductive ducts may be seen on the ventral surface. The **sperm ducts** open by means of prominent pores surrounded by swollen lips on segment 15. The openings of the oviducts on segment 14 and of the seminal receptacles between segments 9 and 10 and 10 and 11 are more difficult to find. Seminal grooves may be visible, running from the sperm ducts in segment 15 back to the clitellum. During copulation, two worms join by their ventral surfaces so that the anterior end of each worm faces the clitellum of the other. Sperm are then exchanged, being discharged by the sperm ducts of each worm, and traveling via the seminal grooves to the seminal receptacles of the other worm. Later eggs from the oviducts and sperm from the seminal receptacles are deposited into the cocoon. Cross-fertilization is thus ensured.

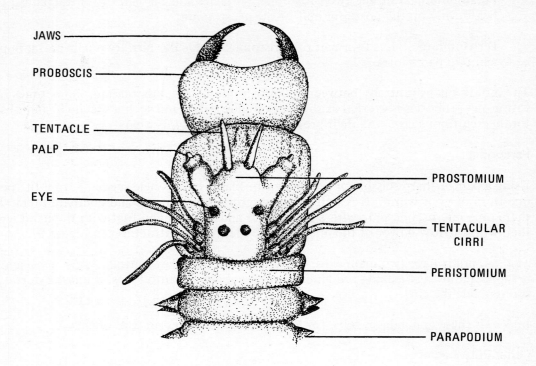

FIGURE 9-1 *NEREIS*, HEAD

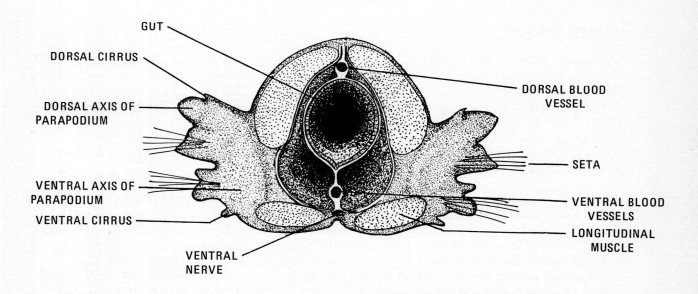

FIGURE 9-2 *NEREIS*, CROSS SECTION

44

Examine the demonstrations of living worms. Note the iridescent **cuticle** which covers the body. It is secreted by the epidermis. It is moist because it secretes mucus and it serves as the respiratory surface for the worm.

B. **Locomotion**

Because muscles can only contract, some provision must be made for returning them to resting length. This is accomplished in the annelids as a result of the fixed volume of the worm. Because water is incompressible, when the circular muscles of a worm contract, the worm must necessarily get longer as well as thinner, thus stretching the longitudinal muscles. Where the coelom is divided by transverse partitions as it is in the earthworm, regions of the body may act relatively independently of one another. In typical earthworm locomotion, waves of contraction pass along the animal. The setae provide traction.

C. **Internal Anatomy**

Pin your specimen dorsal side up in the dissecting pan, putting pins through the prostomium and in a region just posterior to the clitellum. Add just enough cold water to cover the worm. Make a median cut along the **dorsal** midline from the clitellum to the prostomium, making certain that your cut is no deeper than the body wall. If you separate slightly the edges of the cut, you will see that the body wall is attached to the internal organs by means of fine transverse membranes or **septa** (singular: septum). The septa are the partitions that separate the worm into segments and correspond in position to the external grooves of the body wall. The space between the internal organs and the body wall is the body cavity or true coelom.

Using a dissecting needle, carefully tear the septa to separate the internal organs from the body wall. Unroll the body wall on each side and pin it flat, inserting pins obliquely at intervals. Marker pins in segments 5, 10, 15, and 20 will aid in orientation. Refer to Figure 9-3 as you examine the internal anatomy.

1. DIGESTIVE SYSTEM

    The mouth lies under the dorsal lip or prostomium and opens directly into a short buccal cavity which leads to a somewhat swollen muscular **pharynx**, with muscle fibers radiating to the body wall. A rather long, slender **esophagus** follows but is obscured by other organs. Do not remove them at this time. A swollen, soft-walled **crop** or storage chamber and a hard-walled **gizzard** or grinding chamber occur next. The **intestine** extends through the remainder of the worm, and feces are voided through the **anus** in the last segment.

2. REPRODUCTIVE SYSTEM

    The earthworm is monoecious, so a single individual will have reproductive organs of both sexes. The most obvious male structures are the pouchlike **seminal vesicles**. The testes lie inside but are so small that you would have great difficulty in finding them. Sperm ducts, the external openings of which you have already located, drain the seminal vesicles.

    Female organs include the two pairs of **seminal receptacles**, small, almost spherical bodies, located in segments 9 and 10. They open ventrally between segments 9 and 10 and 10 and 11. They store sperm cells received in copulation. A pair of small **ovaries** lie in segment 13 (see the demonstration slide). If you free the intestine carefully, you may possibly be able to see the ovaries with the aid of a hand lens. An oviduct with a funnel-shaped opening lies behind the ovary.

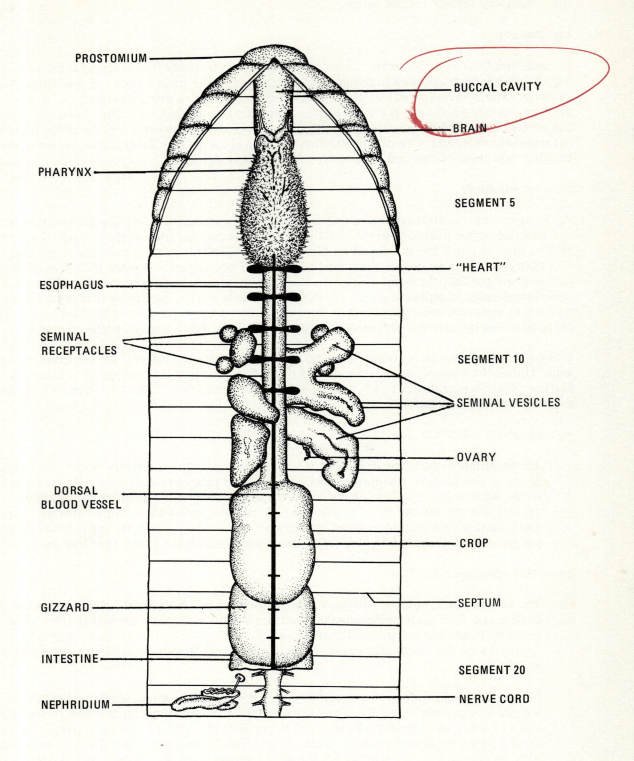

FIGURE 9-3 EARTHWORM, DISSECTION

3. CIRCULATORY SYSTEM

The circulatory system of the earthworm is a closed system of vessels and capillaries containing red blood, red because of hemoglobin contained in the plasma. There is no true heart. Propulsion of blood is accomplished by five pairs of pulsatile branches of the prominent dorsal blood vessel, which is itself pulsatile. The five pairs of pulsatile vessels ("hearts") occur in segments 7 through 11. There are several sets of longitudinal and transverse vessels, some of which may be seen in the slide of cross sections that you will study shortly (see the demonstration of circulation in the living worm).

4. EXCRETORY SYSTEM

The excretory organs consist of paired tubules, known as **nephridia**, occurring in almost every segment. Each nephridium has a funnel-shaped opening in the coelom and a coiled or looped portion, richly supplied with capillaries. The nephridia discharge their wastes through pores on the ventral surface (see the slide on demonstration).

5. NERVOUS SYSTEM

The "brain," composed of a pair of **supraesophageal ganglia**, may be seen on the dorsal anterior surface of the pharynx as a white, bilobed body. A connective from each side passes around the pharynx to the subesophageal ganglion ventrally. The **nerve cord** continues posteriorly for the length of the worm and shows a ganglion or swelling in each segment. Using the hand lens you may see three pairs of transverse nerves given off in each segment.

E. **Cross Section**

A number of sections from two different worms are on the same slide. Those stained red are in serial order through at least one full segment of the worm. Study the sections until you understand the relationships of the parts.

Find the structures in your sections which are labeled in Figure 9-4.

The **typhlosole** is an internal fold of the intestine. It increases the absorptive area of endoderm. Function? How was this same function met in planaria?

Examine the demonstration of nephridia and find sections of nephridia on your slide.

F. **Other Oligochaetes**

Examine the small freshwater oligochaetes on demonstration under the dissecting microscope. What are the prominent annelid characters that permit you to identify the phylum to which these animals belong?

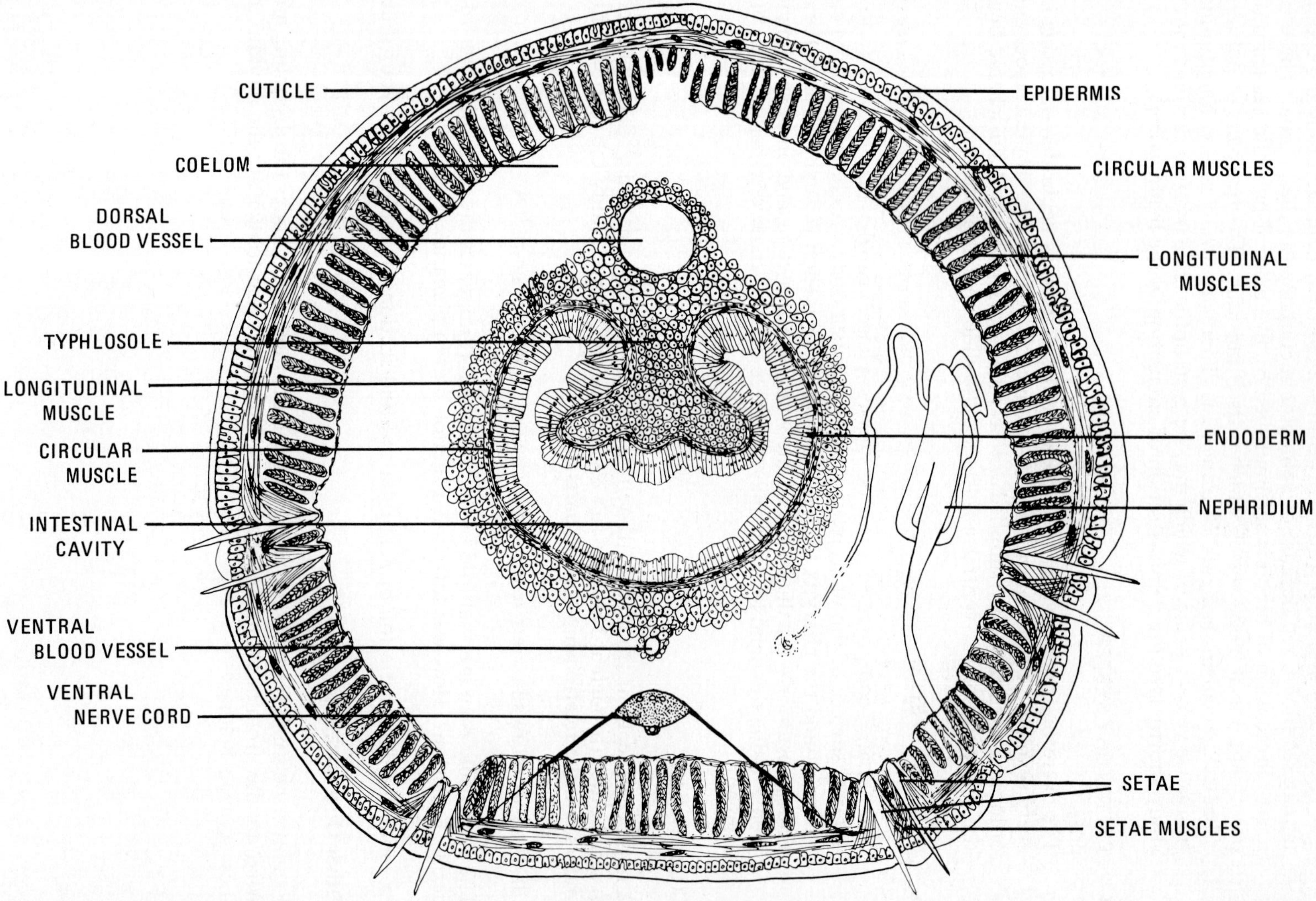

FIGURE 9-4 EARTHWORM, CROSS SECTION

QUESTIONS

1. Using blue for ectoderm, red for mesoderm, and yellow for endoderm, diagram cross sections of a flatworm, a nematode, and an annelid.

2. How is antagonism between longitudinal and circular muscle accomplished in flatworms? What antagonizes the action of the longitudinal muscles of *Ascaris?*

3. Make a table comparing the three worm phyla you have studied with respect to the organ systems present. Which are present in all three? How are the functions represented by new systems in the annelids served in nematodes and platyhelminths?

10. PHYLUM MOLLUSCA

Mollusca (molluscus = soft).

Approximately 100,000 species are known.

Members of the phylum Mollusca live on land, in the sea, and in fresh water. The phylum has the following characteristics (which are secondarily lost in some molluscs).

1. Three body regions.

    a. Head, which is conspicuous in most molluscs other than clams. It bears the mouth and specialized sensory structures.
    b. Foot, which primitively consists of a muscular region of the body wall specialized for the creeping movement that characterizes snails. It may be lost or highly modified; in the cephalopods, for example, it forms tentacles.
    c. Visceral mass, the region of the body that lies dorsal to the foot and contains organ systems.

2. Mantle, the body wall with the exception of the head and the foot. It is specialized to produce the conspicuous shell that is a primitive characteristic of molluscs.

3. Radula, a flexible chitinous ribbon that bears rows of teeth. It is drawn back and forth across cartilaginous bars in the buccal (mouth) cavity and can be protruded to act as a rasp for procuring food. It may be modified and used for such purposes as browsing on encrusted plants or boring through the shell of other molluscs. See the demonstration slide of the radula of a snail.

4. Gills, which were only for respiration in primitive molluscs. Now ciliary-mucoid feeding, in which the gills filter suspended food particles from the current of water that flows over them, is common.

5. Lack of segmentation. In their development, molluscs are startlingly similar to annelids. Cleavage is quite similar, and the same larval form (trochophore) is typical in both phyla. However, molluscs and annelids are strikingly different in that most living molluscs lack segmentation, a conspicuous feature of annelids.

The phylum Mollusca contains six classes.

Class Monoplacophora. Primitive molluscs, such as *Neopilina,* which show internal segmentation.

Class Amphineura. Chitons.

Class Scaphopoda. Tooth shells.

Class Pelecypoda. Clams.

Class Gastropoda. Snails.

Class Cephalopoda. Squids, octopuses.

I. FRESHWATER CLAM

You will not dissect a specimen but rather will examine preparations that make it possible for you to observe the major features of the group.

Examine the clam which has been prepared for each pair of students. One valve of the shell has been removed, together with the ventral portion of the mantle on that side.

A. General Anatomy

Observe the valve that has been removed. Note the attachment of the adductor muscles, which serve to close the valves of the shell. These are antagonized by the hinge ligament, which tends to hold the valves open. Identify the features labeled in Figure 10-1. Notice the general orientation of the animal. The hinge is dorsal in position, the valves lateral, and the foot extends anteroventrally. Refer to the demonstration specimen for internal anatomy.

B. Gills

Notice that there are two pairs of gills. Refer to the diagrammatic cross section of the clam in Figure 10-2. The gills are attached to the mantle and to the visceral mass so as to form a sheet of tissue dividing the ventral infrabranchial chamber completely from the dorsal suprabranchial chamber. Each of the gills is actually a double structure composed of two **lamellae**.

Examine the slide of the cross section of a gill. Each lamella is composed of a large number of filaments. Use Figure 10-2 for orientation. Note the cilia on each of the filaments.

Functionally, the gills serve both for respiration and collecting food. A current of water is drawn into the infrabranchial chamber from the environment and passed to the suprabranchial chamber and thence to the exterior. The pumping force is provided by the lateral cilia, which produce the water flow diagrammed. The long laterofrontal cilia have an effective stroke in the opposite direction and serve to strain out suspended particles and pass them forward to the frontal cilia. Particles are trapped in mucus on the frontal surface of the filaments and passed along toward the mouth by the frontal cilia. Before reaching the mouth, particles are sorted by the **palps**. Observe the demonstration of the trapping of carmine particles by the gill of a living clam.

II. OTHER MOLLUSCS

Observe the demonstrations of members of the other major classes of the molluscs.

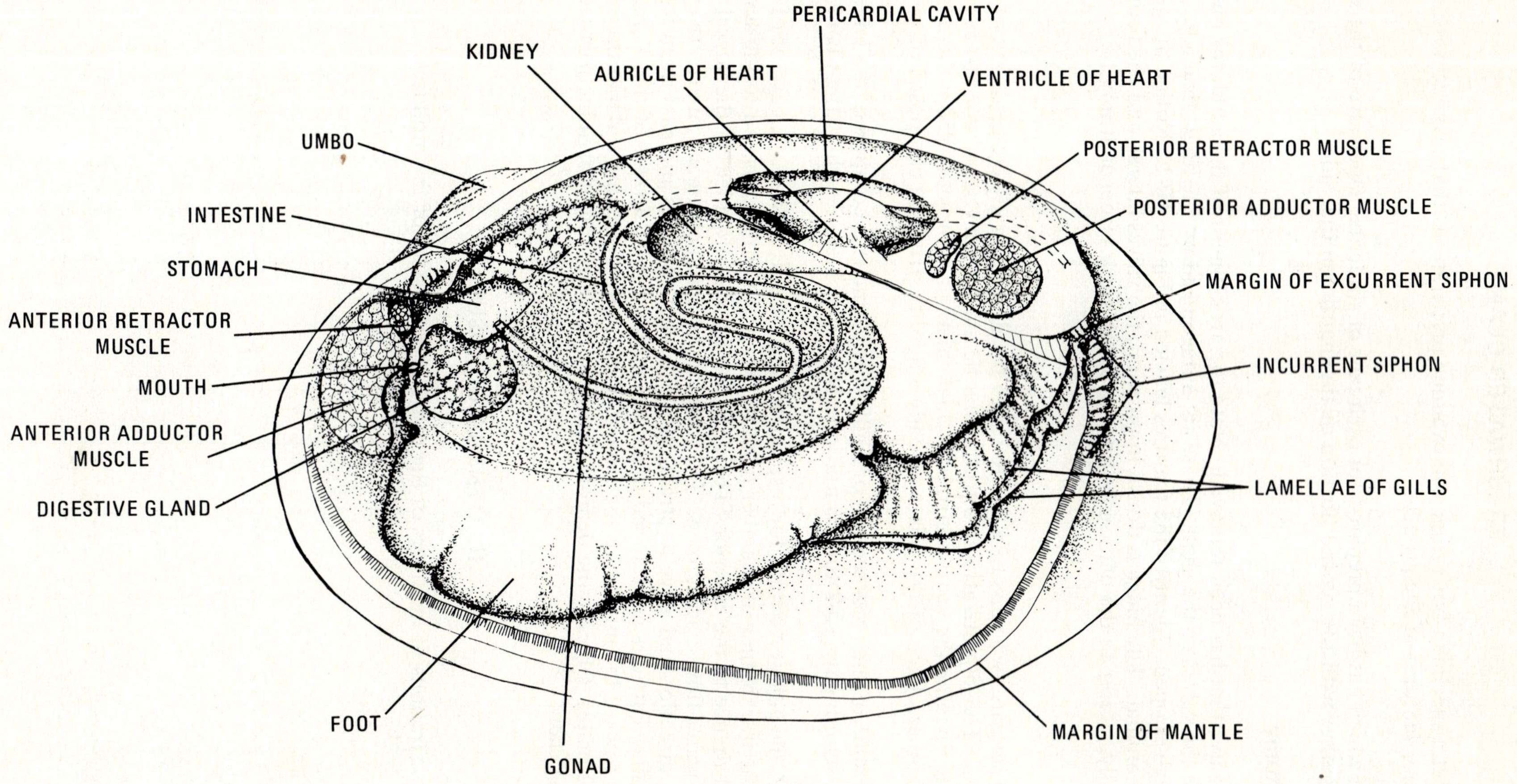

FIGURE 10-1 CLAM, LONGITUDINAL SECTION

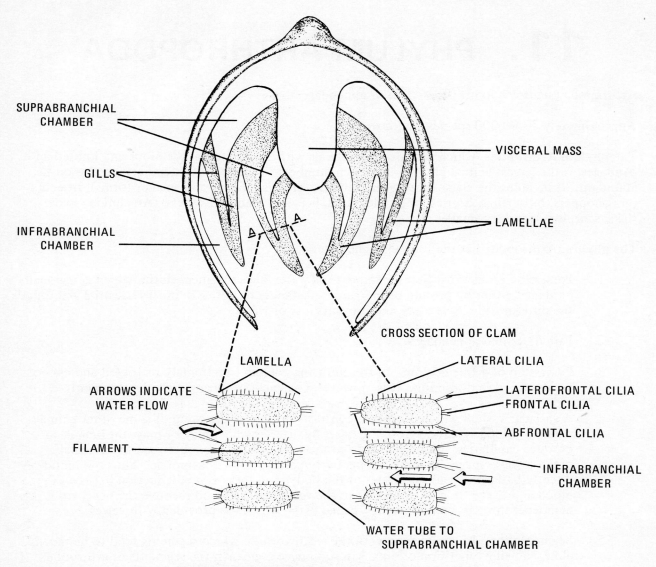

FIGURE 10-2 CLAM, CROSS SECTION AND GILL ACTION

QUESTIONS

1. What are the various functions of larval forms in the life cycle of the animals you have studied?

2. What is the implication of the occurrence of similar larval forms in molluscs and annelids?

3. What information would you need to decide whether it is possible for a clam to obtain sufficient food to survive using the filter feeding apparatus described?

53

11. PHYLUM ARTHROPODA

Arthropoda (arthros = joint, pous = feet or appendages).

Approximately 1,000,000 species are known.

The phylum Arthropoda, which includes the animals with hard external skeletons and jointed appendages, is the largest animal phylum in known numbers of species. Its members are also quite abundant. It includes the classes **Crustacea** (lobsters, crayfish, crabs, and smaller forms), **Insecta** (grasshoppers, butterflies, beetles, bugs, flies, ants, lice, and a host of others), **Arachnida** (spiders, ticks, scorpions), and six smaller classes.

The phylum Arthropoda has the following characteristics.

1. Possession of an exoskeleton composed of a flexible substance, **chitin,** which is regionally hardened either by protein or inorganic substances or both. In most terrestrial arthropods, the outermost layer is waxy and prevents loss of water.

2. Paired, jointed appendages.

3. Possession of a true coelom. In the adult the coelom is extremely restricted and cannot be identified in dissection. It is, however, conspicuous during early development.

4. A vascular system which is **open** in contrast to the closed system of some other animals studied. Blood is carried from the heart by arteries which may or may not branch to produce capillaries. Blood passes from the vessels into spaces between the internal organs. These spaces or sinuses collectively constitute the haemocoel. Thus the blood directly bathes the various tissues of the body, and there is no distinction between the blood and tissue fluid as is the case in the vertebrates. Blood reenters the heart from the hemocoel through small openings, called ostia, which are provided with valves.

5. Segmentation. In comparison with the earthworm in which segments tend to be much alike, showing repetition of organs in successive segments, the segments of arthropods show considerable variation associated with their assumption of specific functions, for example, sensory feeding, defense, locomotion, and reproduction.

In the laboratory you will dissect a crayfish as a representative of the class Crustacea and a grasshopper as an example of the class Insecta. In addition you will examine representatives of some of the other classes. During this study, keep in mind that arthropods represent the culmination of an evolutionary line quite distinct from that of the vertebrates.

I. CRAYFISH

A. **Response of Living Crayfish**

Examine the living crayfish on demonstration. Note its coordinated walking and swimming movements. Tickle an antenna with a dissecting needle. What is the defense response? Rub the surface of the rostrum. What is the response of the eyes? Turn the crayfish over and watch the righting response. If it is disposed to eat, watch its feeding activities. Your instructor will demonstrate water circulation in respiration by placing a drop of India ink at the posterior edge of the gill cover. Where do the ink particles emerge?

B. **External Anatomy**

As in most crustaceans, the head and thorax are fused into a single region, the **cephalothorax**, whereas the **abdomen** is externally segmented and distinct. Each region possesses a number of segments, and each segment a pair of jointed appendages. These appendages all have the same basic biramous (two-branched) plan, as shown by embryological studies, but have been highly modified for various functions. Because each has the same basic structure, they exhibit **serial homology**. The basic structural plan of the appendages is shown most clearly by the abdominal appendages, called swimmerets. Notice that each is composed of a basal portion and two distal portions. See Figures 11-1 and 11-2.

The head of the crayfish is composed of five segments, each represented by a pair of appendages (antennae, antennules, mandibles, and first and second maxillae). The first three segments of the thorax have become associated with the head, and their appendages are modified to assist in food manipulation (first, second, and third maxillipeds). The remaining five segments of the thorax bear the walking legs. Note that the **chelae**, or claws, are included here. The appendages of the abdomen are called **swimmerets**. The first pair of swimmerets is modified for sperm transfer in the male and is reduced in the female. The last pair of appendages, called **uropods**, forms the lateral portions of the tail fan. The terminal segment, the **telson**, bears the anus on the ventral side.

In the male, the sperm ducts open at the bases of the fifth walking legs. In the female, the oviducts open at the bases of the third walking legs. The seminal receptacle of the female is located between the posterior walking legs. Copulation occurs some time before eggs are laid. When laid, the eggs are cemented to the bristles of the swimmerets by a sticky fluid liberated from glands on the ventral surface of the abdomen and uropods. Embryological development occurs while the eggs are cemented to the swimmerets.

Find the mouth. Carefully remove the mouthparts from posterior to anterior. Note the **gill bailer**, a prominent lateral extension of the second maxilla. The gill bailer draws a current of water across the gills from posterior to anterior. The jaws or mandibles work laterally, unlike vertebrate jaws. The first two pair of appendages are sensory in function. Find the excretory pores on the basal segment of the second antennae. Find the stalked **compound eyes** composed of many functional units, the **ommatidia**. Examine the demonstration slides of lobster and insect ommatidia and compare them with the figures in your text.

Note the hard **carapace**, the exoskeleton covering the dorsal side of the thorax. It extends down on each side to form a gill cover. Using scissors, snip off this lateral extension of the carapace on the left side, thus exposing the **gills**. The gill distribution for a typical segment is as follows: one on the first leg joint, two on the soft tissue between the body and the leg, and one on the body wall itself. Four gills per segment is the maximum number; some segments have less.

C. **Internal Anatomy**

Using scissors, cut across the carapace dorsally at the level of the mouth (do not cut ventrally). Extend your cut to include the gill cover on the right side. Carefully work a probe between the carapace and the hypodermis, starting from the posterior border. When it has been freed from the underlying tissue, the carapace can be lifted off exposing the organs of the cephalothorax. Carefully clean away the hypodermis (the epithelium which secretes the exoskeleton). Refer to Figure 11-3 as you dissect.

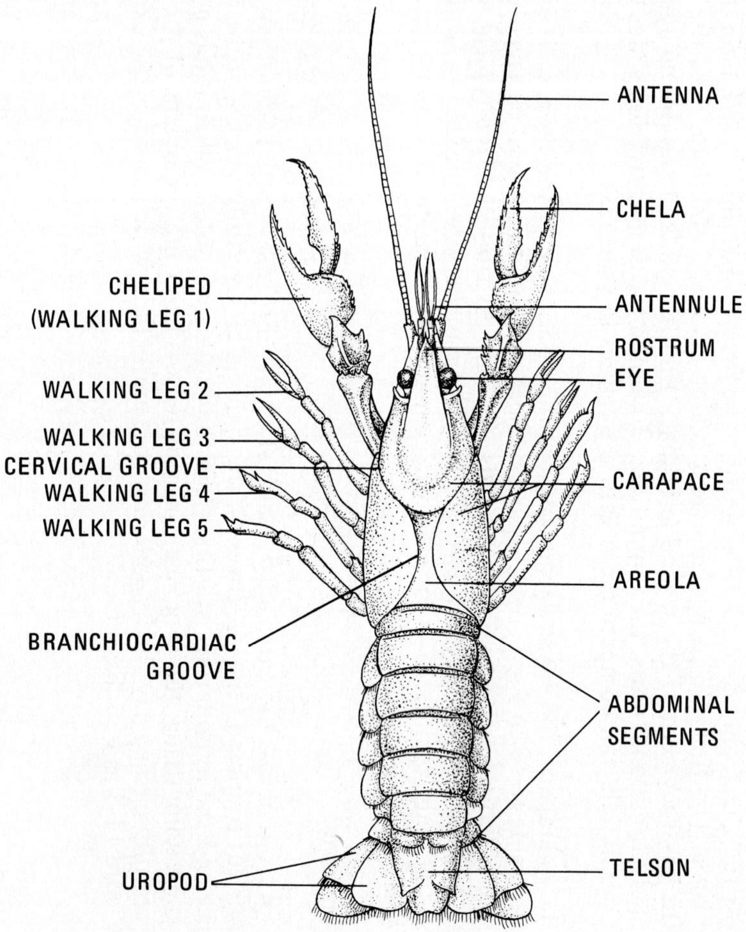

FIGURE 11-1 CRAYFISH, GENERAL DORSAL APPEARANCE

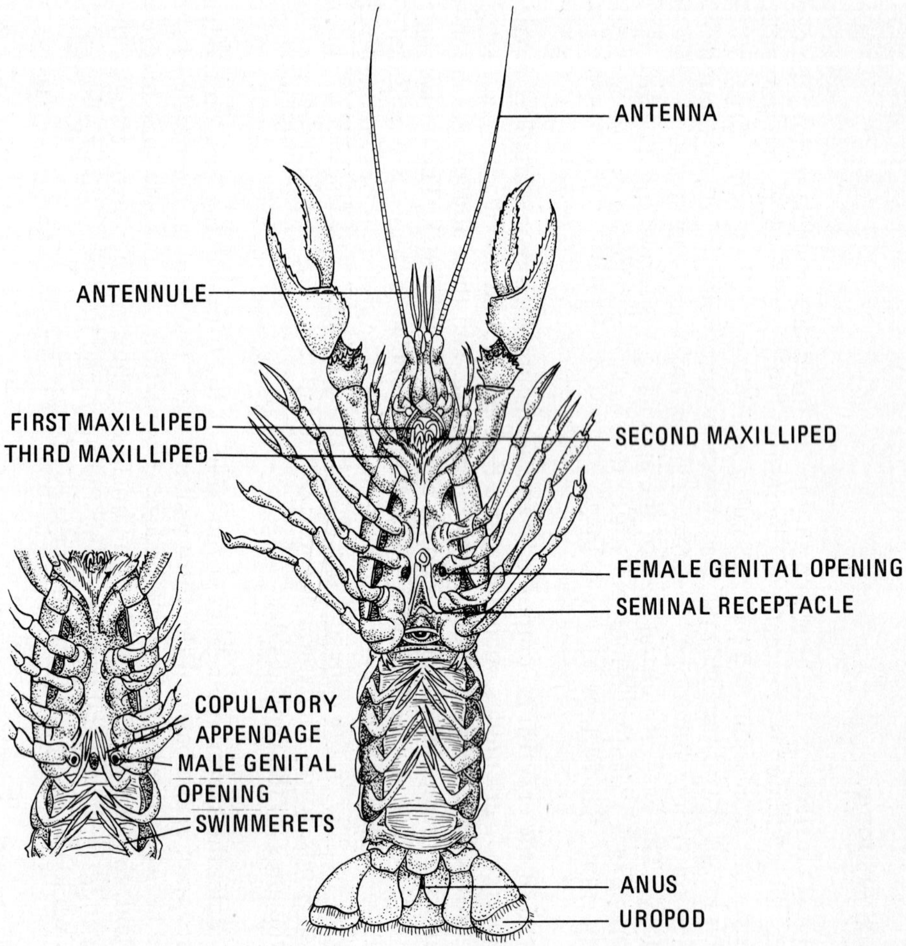

FIGURE 11-2 CRAYFISH, GENERAL VENTRAL APPEARANCE MALE AND FEMALE

1. CIRCULATORY SYSTEM

   It is not practical to attempt a dissection of even the major vessels of the circulatory system unless the specimen has been injected with latex dye. Observe the demonstration dissection of an injected lobster. The heart is quite prominent and may well be beating if your specimen has been freshly chloroformed. The heart lies in the pericardial sinus. Blood enters the heart through three pairs of valved slits, the **ostia**. The arteries leaving the heart branch repeatedly and finally convey the blood to tissue spaces where it bathes the tissue cells directly. These spaces are continuous with larger **sinuses**. Collectively these spaces and sinuses constitute the **haemocoel**. Blood reaches the ventral sinus in the thorax, passes through the gills, and is returned to the pericardial sinus.

2. REPRODUCTIVE SYSTEM

   Remove the heart. Just ventral to it lie the gonads. Both testes and ovary are Y-shaped organs with the single arm of the Y directed posteriorly. The anterior arms run along the sides of the digestive gland. Find the **oviducts** and trace to their external openings. Also follow the course of the **vas deferens** (sperm duct). Near the external openings, the sperm ducts are typically charged with a gelatinous, white fluid containing many sperm. Make a slide of this fluid and see if you can identify the sperm. Unlike the sperm of most animals, these are not flagellated but have several short movable processes.

3. DIGESTIVE SYSTEM

   The digestive system consists of a short **esophagus**, a **stomach**, an **intestine**, and the large, paired, greenish-yellow **digestive glands**. Cut open the stomach from the dorsal side and find the **gastric mill**. It breaks food into small particles. Also notice the prominent strainers. There are usually a pair of hard, white bodies against the side walls of the stomach. These bodies are composed of calcium carbonate, which is stored pending the next moult. They are then used for hardening the new, and at first soft, exoskeleton. The digestive glands are connected by fine ducts to the posterior portion of the stomach. They are compound tubular glands. The digestive glands produce enzymes for digestion (which occurs chiefly in the stomach) and absorb and store digested foods. The intestine has little function except to pass undigestable solids to the anus.

4. EXCRETORY ORGANS

   In the anterior portion of the body just behind the antennae lie the green glands, which remove wastes from the fluid of the haemocoel.

5. NERVOUS SYSTEM

   The double **ventral nerve cord** runs back to the tip of the abdomen beneath the flexor muscles. It may be exposed in the abdomen simply by the removal of these muscles, but in the thorax it is further concealed by internal extensions of the exoskeleton. These extensions should be broken away with a pair of forceps. Note that there is a **ganglion** in each segment. How many abdominal ganglia are there? How many thoracic ganglia? How many nerves form a ganglion? In the head region find the large **subesophageal ganglion**. It is connected by means of a nerve cord on each side of the esophagus to the **supraesophageal ganglion**, or brain.

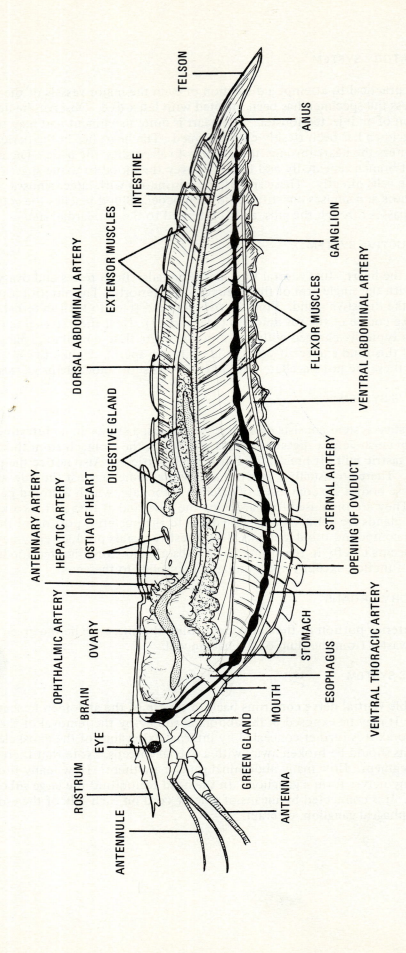

FIGURE 11-3 FEMALE CRAYFISH, LONGITUDINAL SECTION

II. GRASSHOPPER

A. External Anatomy

1. HEAD

    The head bears a pair of **compound eyes**, three simple eyes, or **ocelli** (arranged in a triangle with the apex directed ventrally), a pair of antennae, and several mouthparts (see Figure 11-4). With the aid of Figure 11-5 identify and carefully remove the **labium** and the paired **maxillae** and **mandibles**. Although the **labrum** functions as an upper lip, it does not represent a pair of appendages as do the first three structures mentioned. These three pairs of appendages form the mouthparts of insects in general and are modified to serve a variety of functions. Examine the chart illustrating such modifications. Also examine the demonstration slides of mouthparts specialized for sucking and for piercing and sucking.

2. THORAX

    The thorax has three segments, each bearing a pair of legs. The first segment, the **prothorax**, has a dorsal shield that extends back over the next two segments. The **mesothorax**, or middle thoracic segment, bears a pair of leathery **wings**, and the **metathorax**, the third thoracic segment, has a pair of folding, membranous wings. The mesothorax also bears the largest **spiracle**, into which enters most of the air used in respiration. The legs are comprised of six segments (in the grasshopper, the fifth is divided into three subsegments).

3. ABDOMEN

    The abdomen is not clearly separated from the thorax. The first abdominal segment contains a **spiracle** and, behind it, the large **tympanic membrane**, the "window" to the auditory sac. Other segments also contain tiny spiracles along the lateral borders, through which air is exchanged. The posterior tip of the abdomen differs in the two sexes. That of the female has strong **ovipositors** with which she bores an appropriately deep hole in the ground where the eggs are then deposited. The abdomen of the male has a rounded tip. See Figure 11-4.

B. Internal Anatomy

The dissection should be performed with the specimen submerged in water to avoid drying of the tissues. Using fine scissors, carefully cut through the exoskeleton. This cut should begin at the dorsal midline of the last segment and pass about one-eighth of an inch above the spiracles along either side of the abdomen. Extend these lateral cuts along the side of the thorax to the neck region, but do not cut into the head at this time. Cut through the upper half of the neck region. Now carefully lift the dorsal portion of the body wall, freeing it gradually from underlying muscles, working from the rear forward. Leave as much as possible in the ventral portion of your dissection. Now pin both the dorsal and the ventral portions in a dissecting pan. Be sure they are covered with water.

1. CIRCULATORY SYSTEM

    The circulatory system will remain in the dorsal portion of your dissection. A membrane-like sheet of tissue lies just below the dorsal midline, forming a sinus in which the heart is found. Carefully tease away this membrane and examine the heart. Blood enters through

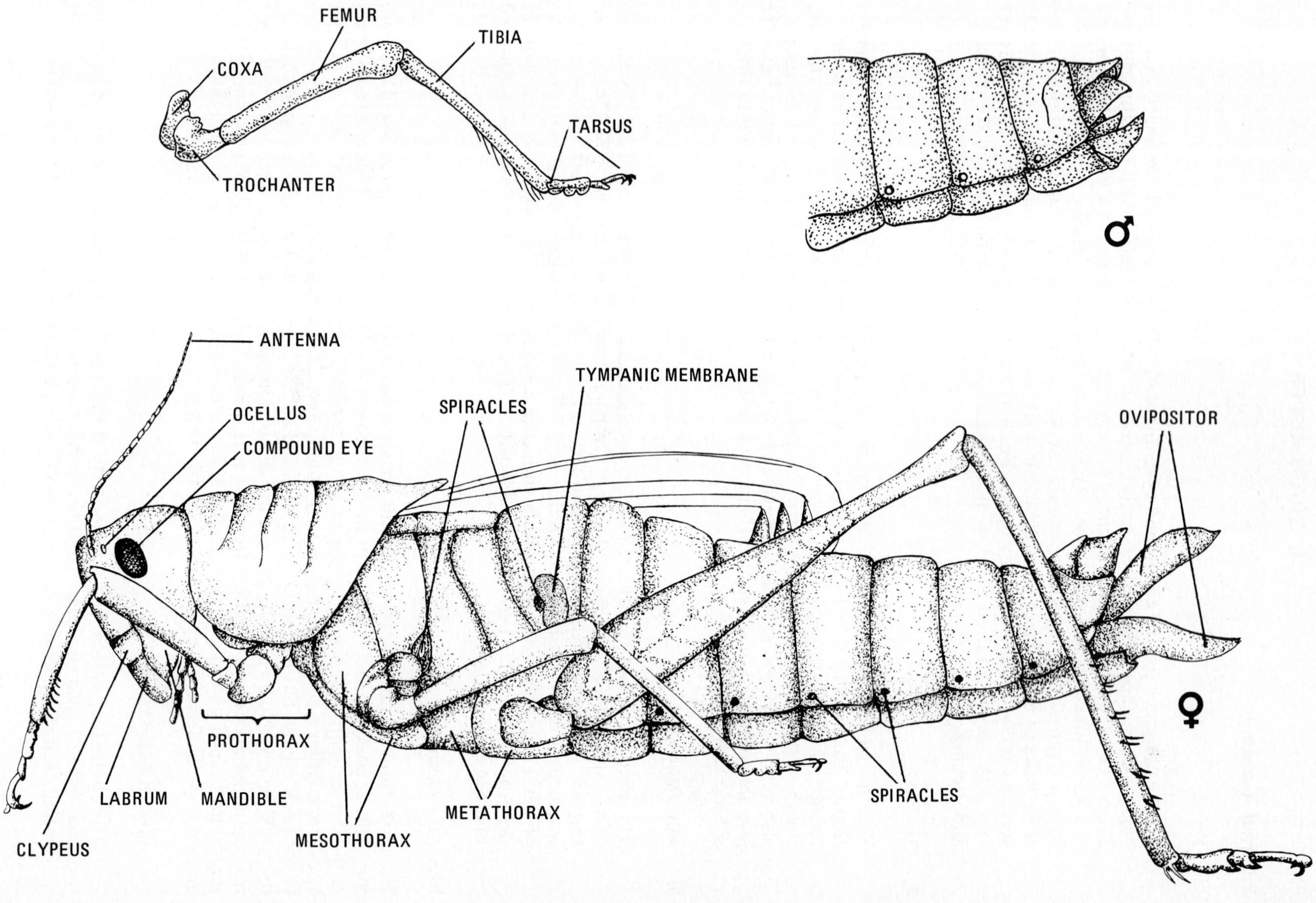

FIGURE 11-4 GRASSHOPPER, EXTERNAL ANATOMY

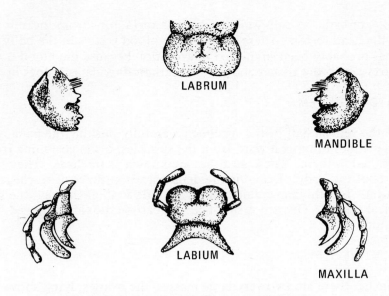

FIGURE 11-5 GRASSHOPPER, MOUTHPARTS

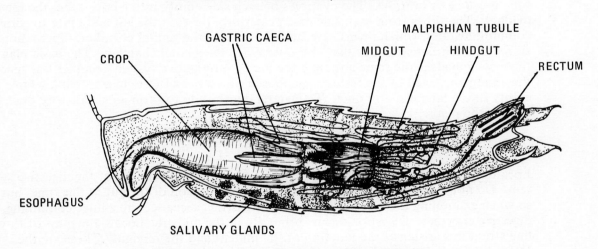

FIGURE 11-6 GRASSHOPPER, DIGESTIVE TRACT

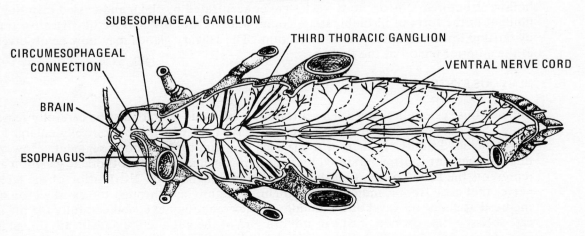

FIGURE 11-7 GRASSHOPPER, NERVOUS SYSTEM

61

paired slits, or **ostia**. A thin-walled tube, the **aorta**, runs from the anterior end of the heart to the head region. There are no other blood vessels in the body. The circulatory system has probably been simplified in evolution because the blood does not function to provide oxygen to the tissues and remove carbon dioxide as it does in most other animals.

2. RESPIRATORY SYSTEM

Examine the ventral half of the specimen, carefully pushing the gonads to one side. Note the slender tubes running inward from the spiracles, often glistening from the air bubbles still contained in them. These air tubes are known as **tracheae**. They ramify and extend to all parts of the body. Examine the demonstration slides of tracheae. In the grasshopper, rhythmic breathing movements provide a flow of air through the main branches of the tracheal system. Oxygen is supplied and carbon dioxide removed from fine branches by diffusion.

3. REPRODUCTIVE SYSTEM

In the female, there are two very large masses, the **ovaries**, lying above the posterior portion of the gut. Each ovary consists of a series of tubes filled with developing eggs; each tube is called an **ovariole**. The ovarioles of each side empty into a large tube, the **lateral oviduct**. The tubes from each side pass posteriorly beneath the gut and unite to form the **common oviduct**, which opens externally by way of the **genital chamber**. Sperm are stored in a tubular structure that also opens into the genital chamber. The basic plan of the male gonads is the same although the male reproductive system occupies less space in the abdomen. Each of the testes is composed of a series of **testicular tubules**, which empty into a lateral sperm duct. The lateral ducts unite to form an ejaculatory duct that ends in the penis.

4. DIGESTIVE SYSTEM

Locate the structures labeled in Figure 11-6, which is a lateral view of the digestive system. Do not attempt to examine the most anterior region of the gut at this time, but do so when examining the nervous system. The gastric caecae secrete a juice containing digestive enzymes. Digestion occurs primarily in the midgut. Locate the mass of threadlike structures arising at the junction of the midgut and the hindgut. These are the **Malpighian tubules,** the chief excretory organs in insects. As the name implies, they are hollow structures. Nitrogenous wastes are concentrated by them and passed into the hindgut in the form of insoluble urate compounds. Carefully remove the digestive tract by cutting through the crop and the rectum. The **salivary glands** may now be observed in the anterior region of the thorax.

5. NERVOUS SYSTEM

The nervous system lies beneath a ventral sheet of tissue which must be carefully teased away. See Figure 11-7. Observe the **ventral nerve cord** which is periodically enlarged to form **ganglia**. Primitively there was one such ganglion for each body segment. However, this arrangement is often modified by fusion of ganglia. The third thoracic ganglion also represents the fusion of the first three abdominal ganglia. The distribution of the lateral nerves that arise from it indicates this origin. Carefully cut through the exoskeleton of the head at about the level of the middle of the compound eye, and carefully lift off the cap of tissue thus formed. Lift the anterior end of the gut and locate beneath it a large mass of nervous tissue, the **subesophageal ganglion**. Nerves from this ganglion innervate the mouthparts. Arising from the subesophageal ganglion is a pair of large nerves that pass upward around the esophagus, the **circumesophageal connectives.** These lead to the **brain,** which is composed of three pairs of lobes. Nerves from the brain lead to the eyes and the antennae.

III. INSECT METAMORPHOSIS

The **change in form** that takes place between the **larval stage** (the individual which hatches from the egg) and the adult stage (the stage which is sexually mature) is called **metamorphosis.** The change may be gradual as in the grasshopper, or may be relatively sudden as in the butterfly. Gradual metamorphosis is termed **incomplete,** whereas **complete** metamorphosis refers to the abrupt transition to adult form. All intermediate degrees of metamorphosis are found in insects.

A. Examine the glass mount of the **stages in the life history** of the **mourning cloak butterfly.** The **caterpillar** molts four times during growth, leaving cast-off skins, before it becomes quiescent and forms the **pupa.** A complete metamorphosis occurs within the pupa case, and in about 2 weeks the **adult** emerges, leaving an empty shell split along the back, the line of emergence.

B. Examine the **naiads** of the **dragonfly** and the **nymphs** of the grasshopper, which are to be found in the white dishes on your table. Note the partially developed wing pads on both, and the elaborate mouthparts of the dragonfly naiad. The grasshopper is vegetarian; the dragonfly is carnivorous, eating mostly other insects. The mouthparts are modified for capturing and holding the prey.

IV. ORDERS OF INSECTS

A list that characterizes 15 of the common orders of the class Insecta follows. You are expected to become familiar with these orders and to recognize typical examples. You can obtain this information through the laboratory demonstrations, your text, and by using the synoptic collection.

A. **Subclass Apterygota** (without + wings). Primitively wingless.

   1. ORDER COLLEMBOLA (glue + wedge).

      a. **Examples:** Springtails; abdominal segments reduced to six, the first segment bears a ventral tube, the fourth bears a forked springing organ, and the third is equipped with a triggerlike catch. These modified segments enable the little beast to literally spring by means of the "tail." A few species are injurious to plants but most feed upon disintegrated organic matter.

      b. **Characters:**
      1. Primitively wingless.
      2. Chewing mouthparts.
      3. No metamorphosis.
      4. Small and soft-bodied.
      5. Abdominal segments reduced to six.

      c. **Importance:** Abundant in forest soils, may damage young vegetables and mushrooms. There are 2,400 species.

63

2. ORDER THYSANURA (bristle + tail). So named because of two or three long antenna-like tails at the tip of the abdomen.

   a. **Example:** Silverfish, a common household insect, often found in piles of old papers or magazines where it feeds on the starch in paste.

   b. **Characters:**
      1. Primitively wingless.
      2. Chewing mouthparts.
      3. No metamorphosis.
      4. Small and soft bodied.
      5. Rudimentary abdominal legs.

   c. **Importance:** Show affinities with both centipedes and winged insects, thus constituting a "missing link" between lower arthropods and the higher insects. There are 500 species.

B. **Subclass Pterygota** (with + wings). Many members have secondarily lost their wings; for example, Anoplura and Siphonaptera.

   Division 1. Hemimetabola (gradual metamorphosis). Exopterygota (wings develop externally).

3. ORDER ODONATA (a tooth).

   a. **Examples:** Dragonflies, damselflies.
      Suborder Anisoptera. Dragonflies, wings held outward when resting.
      Suborder Zygoptera. Damselflies, wings held erect when resting.

   b. **Characters:**
      1. Two pairs of membranous wings, abdomen slender and elongate.
      2. Chewing mouthparts, adults strong flying predators on other insects.
      3. Gradual metamorphosis, young stages naiads and aquatic, spend 3 months to 5 years in water. There are 5,000 species.

4. ORDER EPHEMEROPTERA (ephemeral + wing).

   a. **Example:** Mayflies.

   b. **Characters:**
      1. Two pairs of membranous wings, held erect when resting.
      2. Chewing mouthparts, vestigial in adult.
      3. Gradual metamorphosis, young stages naiads and aquatic, spend a few months to 3 years in water. There are 1,500 species.

5. ORDER ORTHOPTERA (straight + wing).

   a. **Examples:** Cockroaches, grasshoppers, true locusts, crickets, praying mantis, walking sticks.

   b. **Characters:**
      1. Two pairs of wings, forewings more or less thickened, the hind wings often folded in pleats like a fan.

2. Chewing mouthparts.
3. Gradual or incomplete metamorphosis.

c. **Importance:** Destroy crops, inhabit houses, eat other insects. There are 23,000 species.

6. ORDER PLECOPTERA (to fold + wing).

   a. **Example:** Stoneflies.

   b. **Characters:**
   1. Two pairs of membranous wings, held pleated but flat over the back when at rest.
   2. Chewing mouthparts, often absent in adults.
   3. Gradual metamorphosis, adults weak fliers, young stages aquatic naiads with tufted tracheal gills. There are 1,500 species.

7. ORDER ISOPTERA (equal + wing).

   a. **Examples:** Termites.

   b. **Characters:**
   1. Two pairs of membranous wings or wingless.
   2. Chewing mouthparts.
   3. Gradual or incomplete metamorphosis.
   4. Social insects with at least three castes: the reproductive caste, the worker caste, and the soldier caste. Each caste includes both males and females.

   c. **Importance:** Economically important because some destroy wood products. There are more than 2,000 species.

8. ORDER ANOPLURA (unarmed + tail).

   a. **Examples:** Sucking lice; crab louse, body louse (cootie), head louse of humans.

   b. **Characters:**
   1. Wingless.
   2. Body depressed, flat.
   3. Sucking mouthparts.
   4. Metamorphosis absent or slight.

   c. **Importance:** Ectoparasites of birds and mammals, feed on blood and transmit various diseases, including typhus and European relapsing fever. There are 250 species.

9. ORDER HEMIPTERA (half + wing).

   a. **Examples:** True bugs, electric light bug or giant water bug, water boatman, water strider.

   b. **Characters:**
   1. Two pairs of wings, forewings thick and horny at bases, membranous behind, crossed at rest, hind wings membranous, fold under forewings.
   2. Piercing-sucking mouthparts.
   3. Gradual metamorphosis.

c. **Importance**: Common bedbugs of humans, destroy crops, certain species suck blood from mammals, including humans, species of *Triatoma* often infected with the flagellated protozoan *Trypanosoma cruzi,* which causes Chagas' disease in South America. There are 40,000 species.

10. ORDER HOMOPTERA (like + wing).

   a. **Examples**: True cicadas, aphids, scale insects.

   b. **Characters**:
      1. Two pairs of membranous wings or none, roofed over abdomen when at rest.
      2. Piercing-sucking mouthparts.
      3. Gradual metamorphosis.

   c. **Importance**: Destructive to plants, some species transmit diseases. Cicadas or "locusts," so-called "17-year locusts," spend 13 or 17 years in ground as nymphs. There are 32,000 species.

Division 2. Holometabola (complete metamorphosis). Endopterygota (wings develop internally).

11. ORDER LEPIDOPTERA (scale + wing).

   a. **Examples**: Butterflies and moths.

   b. **Characters**:
      1. Two pairs of wings covered with scales arranged like the shingles on a roof. Scales are modified hairs.
      2. Sucking mouthparts in adults, chewing mouthparts in larvae.
      3. Complete metamorphosis.

   c. **Importance**: Clothes moths, forest tent caterpillars, European corn borer, silkworm moth, many species play important roles in cross-pollination of flowers. There are 112,000 species.

12. ORDER DIPTERA (two + wing).

   a. **Examples**: True flies, mosquitos, midges, gnats, crane flies, horseflies, houseflies, fruit flies.

   b. **Characters**:
      1. Forewings transparent, hind wings represented by short knobbed halters, some species wingless.
      2. Piercing-sucking or "sponging" mouthparts.
      3. Complete metamorphosis.

   c. **Importance**: Mosquitos transmit or serve as vectors for various diseases, *Culex* transmits bird malaria, *Anopheles* transmits *Plasmodium* of humans, certain species, such as the botfly, are parasitic in the larval stage; other species feed on domestic plants ranging from chrysanthemums to wheat. *Drosophila* spp., called fruit flies, have importance in genetics. The common housefly may contaminate food and spread disease. There are 80,000 species.

13. ORDER SIPHONAPTERA (tube + without wing).

   a. **Example:** Fleas.

   b. **Characters:**
      1. Wingless.
      2. Body strongly compressed laterally.
      3. Sucking mouthparts.
      4. Complete metamorphosis.

   c. **Importance:** Ectoparasitic on birds and mammals, jigger on feet of humans and other mammals, Indian rat flea transmits organism that causes bubonic plague. There are 1,100 species.

14. ORDER COLEOPTERA (sheath + wing).

   a. **Examples:** Beetles and weevils, including such forms as the potato beetle (bug), June beetle (bug), "fire flies," tiger beetles, ladybird beetles.

   b. **Characters:**
      1. Two pairs of wings, forewings (elytra) thick and leathery, veinless, meeting along middorsal line, hind wings membranous, fold forward under forewings when at rest. Some species wingless.
      2. Chewing mouthparts.
      3. Complete metamorphosis.

   c. **Importance:** Very destructive of crops, wood borers, some species beneficial in that they prey on aphids and scale insects. There are more than 280,000 species.

15. ORDER HYMENOPTERA (membrane + wing).

   a. **Examples:** Wasps, hornets, bees, ants.

   b. **Characters:**
      1. Two pairs of wings or none. Wings small, membranous, interlocking in flight.
      2. Chewing mouthparts or chewing-lapping mouthparts.
      3. Complete metamorphosis.

   c. **Importance:** Certain species destructive of trees and other plants, larvae of other species are parasitic on other destructive insects, honey bees, ants, bees with interesting caste systems and social organization. There are 105,000 species.

V. OTHER ARTHROPODS

Representatives of several major groups of the arthropods will be found at your table. In addition to the insects and crustaceans, this phylum also includes such animals as spiders, ticks, mites, centipedes, and millipedes.

It is possible to place arthropods in their proper class in the group by referring to a few simple characters. The following **key** will permit you to classify your specimens. This key is termed **dichotomous** because it is composed of paired statements. At each point, a simple choice must be made. This choice in turn designates another choice until the identification is finally made.

1. One or more pairs of antennae 3
   No antennae 2

2. Five or six pairs of abdominal appendages modified as gills and
      a spikelike telson . . . . Merostomata
   Abdomen without appendages and lacking a spikelike telson . . . . Arachnida

3. One pair of antennae 4
   Two pairs of antennae . . . . Crustacea

4. Body consists of a head and long trunk region 5
   Body of head, thorax, abdomen, three pairs of legs . . . . Insecta

5. One pair of legs per segment . . . . Chilopoda
   Two pairs of legs per segment . . . . Diplopoda

Place your specimens in their respective classes. Construct a dichotomous key to four or five specimens using characteristics other than those used above.

A. Class Crustacea

Crustaceans exhibit a considerable variety of form. You have already examined a crayfish in some detail and have seen several crabs. Examine living specimens of *Daphnia* (the water flea), *Cyclops,* and the brine shrimp, *Artemia.* Can you identify two pairs of antennae in each of these?

B. Class Arachnida

This group includes the spiders, ticks, and mites. There may be either two body regions, as in the spiders, or only one (ticks and mites). There are six pairs of appendages. Mandibles are not present but are replaced by **chelicerae,** which may either be clawlike or modified as fangs. The second pair of appendages, the **pedipalps,** are sensory in function.

C. Class Merostomata

Examine the horseshoe crab, *Limulus.* Note that despite its common name, its characteristics are similar to those of the arachnids (though the chelicerae and pedipalps are much like the walking legs).

D. Classes Chilopoda and Diplopoda

Both of these classes are mandibulate (bear mandibles rather than chelicerae). Centipedes (class Chilopoda) have the first pair of legs modified as poison claws. The bite of even the small house centipede is rather painful. Millipedes (class Diplopoda) are vegetable scavengers in contrast to the predacious centipedes. They have been described to liberate poison gas to discourage predators.

VI. PHYLUM ONYCHOPHORA

This is a small group of extremely restricted distribution, but it is important to our study because of its relationships both to the Annelida and the Arthropoda. It includes the genus *Peripatus,* an organism sharing typical annelid and arthropod characteristics. It is one of the few living "connecting links" between major animal groups. It is sometimes considered a class of the Arthropoda, but is more commonly classified as a separate phylum.

Compare the specimens of *Peripatus* with the annelids and with the centipedes and millipedes. Notice that the legs are segmented but not jointed. There is no exoskeleton. Internally, there is a coelom and nephridia. It is likely that *Peripatus* is related to the evolutionary stem which, diverging from the ancient annelids, gave rise to the primitive arthropods.

QUESTIONS

1. What characteristics would you use to assign each of the following animals to its proper class: crayfish, grasshopper, spider, and centipede.

2. Briefly describe the circulatory, respiratory, excretory, and nervous systems of a typical insect such as a grasshopper.

3. In what ways are insects important to us?

4. Why do you think arthropods are so successful?

12. PHYLUM ECHINODERMATA

Echinodermata (echinos = hedgehog or sea urchin, derma = skin).

Approximately 6,000 species are known.

The phylum Echinodermata is exclusively marine. Most echinoderms, such as starfishes, are slow-moving; others, such as sea lilies, are attached and superficially resemble plants more than animals. Echinoderms have three germ layers and a true coelom. They are not segmented.

The phylum Echinodermata has the following characteristics.

1. Radial symmetry

    The body of most echinoderms is radially symmetrical with five or some multiple of five radial divisions. Almost always there are minor departures from strict radial symmetry in either external or internal anatomy, showing that this condition is secondary. This fact together with the presence of larval forms that are bilaterally symmetrical distinguishes the Echinodermata sharply from such truly radially symmetrical phyla as the Coelenterata. However, because the adult is typically organized about a central axis, it is appropriate to speak of an oral and aboral surface, rather than dorsal and ventral. Correlated with this secondary assumption of radial symmetry is the absence of cephalization (head formation). Thus in the starfish, no arm is physiologically more important than another. The nervous system is correspondingly reduced and consists of strands of tissue following the arms and a rather diffuse superficial network.

2. Water-vascular, or ambulacral, system

    This system is composed of a set of tubes derived from the extensive coelom, typically open to the exterior by way of the **madreporite**. In many echinoderms, the ambulacral system includes tube feet and ampullae which may serve for locomotion, respiration, or feeding.

3. Endoskeleton

    The skeleton of echinoderms is composed of dermal plates or ossicles derived from mesoderm. Also characteristic are the spines which give the phylum its name. The dermal plates vary considerably in the different classes of the phylum, being firmly cemented together to produce the test in sea urchins and surviving only as scattered plates of complicated form in the muscular body wall of the sea cucumbers.

The phylum Echinodermata contains five classes.

Class Crinoidea. Sea lilies.

Class Asteroidea. Starfish.

Class Echinoidea. Sea urchins (refer to *urchin* in the dictionary).

Class Holothuroidea. Sea cucumbers.

Class Ophiuroidea. Brittle stars.

I. STARFISH

A. External Anatomy

Note the central disc, the five radial arms, the **aboral** (opposite mouth) side, and the **oral** side of your starfish. The **madreporite** is found on the aboral side between two of the arms. Notice the bony **dermal spines**. These, together with other bony plates, form the **endoskeleton**. Examine the demonstrations of the **dermal branchiae** and the **pedicellariae**. The latter are small clawlike structures that function to remove debris from the surface of the starfish. How are they arranged with respect to the dermal spines? Sketch one of them in the space provided. Eyespots are located at the tip of each arm, but will not be visible in the preserved specimens.

Turn your starfish over and examine the oral surface. Note the central mouth and the ambulacral groove running along each arm. The tube feet will be visible along each side of each ambulacral groove.

Pedicellaria

B. Internal Anatomy

Dissect open your starfish from the aboral surface by the following procedure: snip off the last inch of the arm **opposite** the madreporite. Cut along each side of this arm to the central disc and then cut around the central disc. Make your cut inside the region of the madreporite so as not to destroy its internal connections. Now gently lift the aboral surface of the cut arm. Note that the internal organs are fastened by mesenteries to the aboral body wall. Carefully tear these mesenteries as you remove the aboral wall so that the organs remain in the coelom. As you remove the aboral wall from the central disc, you will tear the **pyloric stomach**. It is connected by means of a short intestine (which you probably will not see) to the anus on the aboral surface. Identify the structures indicated in Figure 12-1. What is the function of each? Find the strong **retractor muscles** attached to the cardiac stomach. Find the rather brittle tube (the stone canal), which connects the madreporite to the ring canal. The **ring canal** is located within the bony plates surrounding the mouth. In addition to its connection with the stone canal, it is connected with a **radial canal** in each arm. These in turn are connected by small lateral canals to each tube foot. These canals, together with the tube feet and ampullae, form the water-vascular system. How do the tube feet operate? What function can you assign to the madreporite?

II. OTHER ECHINODERMS

Examine representatives of the classes of echinoderms on demonstration and learn to identify the class to which each specimen belongs.

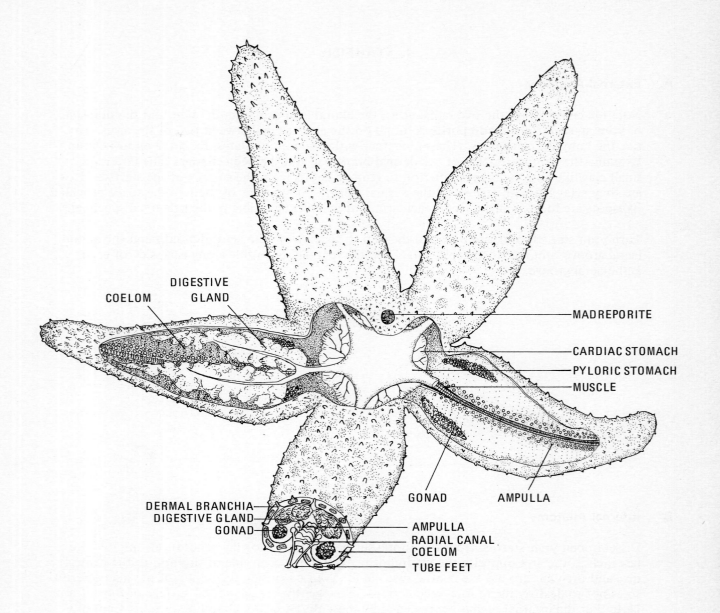

FIGURE 12-1 STARFISH, DISSECTION

III. EMBRYOLOGY AND DEVELOPMENT

Examine the slide of early cleavage stages in the starfish. Unfertilized eggs can be recognized by the prominent nucleus which is no longer apparent in these slides once the egg has been fertilized.

Find 1-, 2-, and 4-cell stages. Study an 8-cell stage and note that the three cleavage planes which produce these 8 cells are at right angles one to another and that the cells are approximately equal in size. This pattern of cleavage is termed *radial* in contrast to the *spiral* pattern of cleavage characteristic of annelids and molluscs.

In many respects, the development of echinoderms and that of chordates are quite comparable. Notably:

1. Cleavage is radial in both groups.

2. Mesoderm arises as paired outpocketings from the **archenteron,** or primitive gut. Examine the demonstration of the starfish gastrula.

3. The mouth does not form at the site of invagination of the archenteron.

For these reasons, the line of eucoelomate (true coelom) animals, which includes the echinoderms and the chordates as well as some smaller groups, is separated from the annelid-mollusc-arthropod line.

Examine the demonstration slides of the **bipinnaria** and **brachiolaria** larvae. Note that both are bilaterally symmetrical. During development, the embryo develops into the free-swimming, feeding bipinnaria larva that subsequently differentiates into a brachiolaria larva. The latter develops into a radially symmetrical miniature adult after further growth.

IV. RELATION OF ECHINODERMS TO CHORDATES

Superficially there is very little resemblance between echinoderms and the simplest and most primitive of the chordates. However, most zoologists believe that these animals are more closely related to us than are the annelids or arthropods. Rather basic similarities in development have already been mentioned. A number of other similarities have been studied more recently. By themselves, these similarities are no more than suggestive. Collectively, and added to the considerable similarities in embryology of the two groups, they strengthen the argument for the relationship.

A. The larval stages of some of the echinoderms are almost identical to those of the Hemichordata, a small phylum of animals whose morphology is more directly comparable to that of primitive chordates.

B. Adult echinoderms possess a mesodermal endoskeleton which is unique for invertebrates but which is also present in chordates and is especially well developed in vertebrates.

C. The proteins of the echinoderms seem to be structurally more similar to the proteins of chordates than are those of any other invertebrate group. This similarity was determined by inducing the formation of antibodies to protein extracts and then determining the strength of cross-reaction of these antibodies to proteins from other groups.

QUESTIONS

1. What does the water-vascular system do? How does it work?

2. What evidence is there that echinoderms and chordates are related?

3. What is a reasonable explanation for the reversion to radial symmetry in adult echinoderms?

13. PHYLUM CHORDATA

Chordata (chord = cord, -ata = characterized by).

Approximately 45,000 species are known.

The phylum Chordata contains most of the familiar animals from fishes to humans and also some less familiar ones. The phylum has the following characteristics.

1. Pharyngeal gill slits at some stage of development.

2. An axial rod of cells, the notochord, as a primitive skeletal axis.

3. A dorsal, hollow, central nerve tube.

The phylum Chordata contains three subphyla.

Subphylum Urochordata. Sea squirts. These are marine organisms, typically sessile and filter feeding. Only the gill slits indicate their chordate affinities as adults, but the larvae exhibit the notochord and hollow dorsal nerve cord.

Subphylum Cephalochordata. Amphioxus and related forms. Adult cephalochordates exhibit all of the chordate features. However, the notochord runs the full length of the animal, extending into the head (hence the name of the group), and other characteristic features of the vertebrates are not present.

Subphylum Vertebrata. Chordates with a vertebral column, including the lampreys, cartilaginous and bony fishes, amphibians, reptiles, birds, and mammals.

The Urochordata and Cephalochordata are often referred to as the protochordates. This association is a marriage of convenience rather than a reflection of close affinities. They are similar in respects that are somewhat negative:

1. Neither has a vertebral column.

2. Neither has prominent cephalization.

3. The gill slits do not exit to the surrounding environment directly but rather empty into an atrial cavity.

As will become apparent, the cephalochordates and urochordates differ in many respects. Hence they will be considered separately.

I. SUBPHYLUM UROCHORDATA

Adult urochordates offer a considerable variety of form. Some are colonial and others solitary. The most primitive class in this group is the class Ascidiacea, the sea squirts. Examine the demonstration of an adult sea squirt while referring to Figure 13-1.

The adult animal is a filter feeder, removing suspended particles from the water pumped through the gill slits by cilia. Note that after passing through the gill slits, water is returned to the environment by first passing into an atrial cavity surrounding the pharynx and thence via an excurrent atrial siphon. Notice that these animals are not segmented as are the other chordates. Adults of the other classes of the urochordates are free-swimming animals, either retaining the larval tail or using the excurrent siphon for propulsion.

The larva of the ascidians, the "tadpole" larva, exhibits the chordate-distinguishing characters in diagrammatic form (Figure 13-1). Find the notochord, gill slits, and nerve cord. Verify the relations of the gill slits and the atrial cavity.

The larva serves primarily in selection of the habitat of the adult sea squirt. It is liberated and typically has a short free-swimming existence. When ready to settle, it responds negatively to light and swims toward shadow. This positions the larva against wharf pilings, under overhanging ledges of rock, and in such places. It then attaches, resorbs the tail, and develops to the adult form.

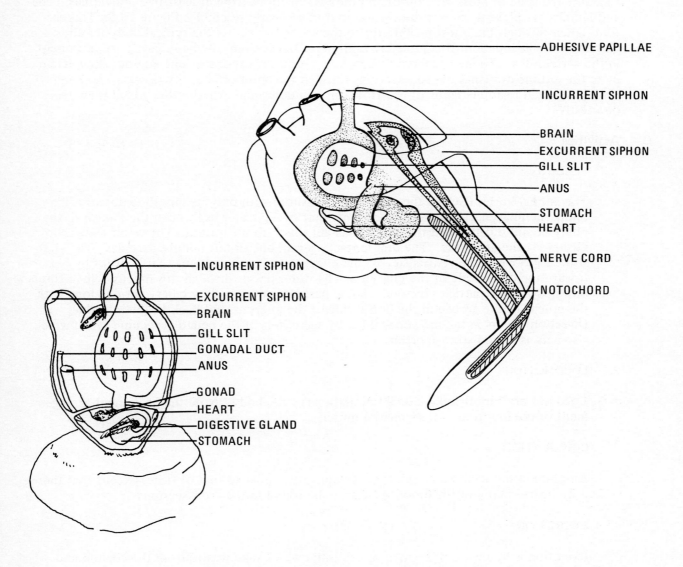

FIGURE 13-1 UROCHORDATE, ADULT AND LARVA

II. SUBPHYLUM CEPHALOCHORDATA

The cephalochordates, like the urochordates, are filter-feeding animals. Amphioxus is typical of the small, fishlike cephalochordates.

A. Whole Mounts

Examine the whole mounts of amphioxus. Identify the structures indicated in Figure 13-2. The description which follows will allow you to understand the function of each organ. Also examine the specimens in dishes on your table. Find the whitish, segmental **gonads** in the central one-third of the body (gonads are not visible in the whole mounts). The small pigment spots along the nerve tube serve as light receptors; the "eyespot" does not.

B. Cross Sections

Examine the slide of cross sections of amphioxus. Sections from at least two individuals, male and female, are present. Select the section that most nearly resembles Figure 13-5. This section, taken through the intestine anterior to the atriopore, shows the typical body features. Note that the atrium almost completely obliterates the coelom, the body cavity so prominent in the vertebrates. Having mastered this section, study the other sections on your slide. Complete the outline drawings of cross sections through the anterior pharynx (Figure 13-3) and through the tail region (Figure 13-6). How can you determine whether the gonad is an ovary or a testis?

C. Activities

1. FEEDING

    The animal buries itself in the sand of the seashore with only its mouth exposed. Cilia lining the pharyngeal bars create currents that sweep water in through the mouth to the pharynx and out through the pharyngeal gill slits into the atrium. The water leaves the atrium via the atriopore. The oral tentacles strain out all but minute particles. The **endostyle**, a groove in the ventral side of the pharynx, secretes a mucus that is swept by cilia toward the dorsal side of the pharynx. The water passes through this mucus, which filters out microscopic particles of food. These particles are passed by ciliary action along with the mucus to the groove in the dorsal side of the pharynx and back to the intestine. Digestion occurs in the intestine, aided by secretions of the **hepatic caecum** (liver), which also acts as a food storage organ.

2. LOCOMOTION

    Locomotion is by fishlike lateral undulations of the body. Note the prominent arrangement of the myotomes or segmental muscles.

3. CIRCULATION

    Amphioxus has a closed circulatory system very similar to that of fishes except that there is no heart. Many of the blood vessels can be found in the cross sections.

4. EXCRETION

    Excretion is by means of segmental nephridia which take wastes from the coelom and empty them into the atrium.

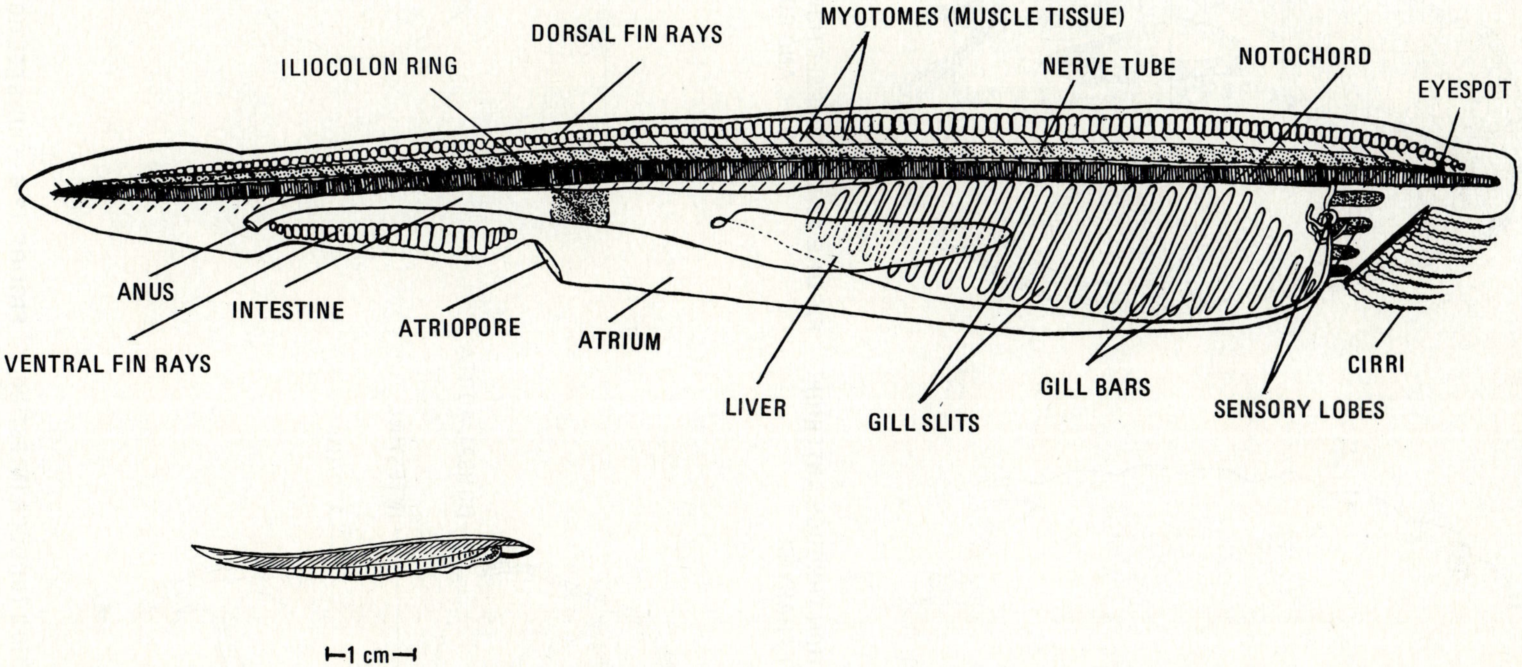

FIGURE 13-2 AMPHIOXUS, GENERAL APPEARANCE AND ANATOMY

FIGURE 13-3 CROSS SECTION THROUGH ANTERIOR PORTION OF PHARYNX

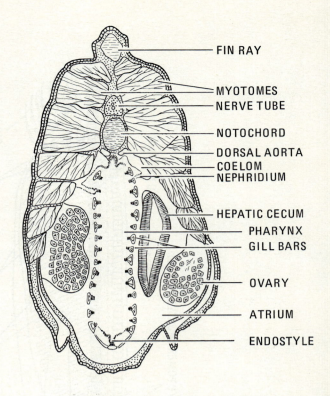

FIGURE 13-4 CROSS SECTION THROUGH PHARYNX, LIVER, AND GONAD

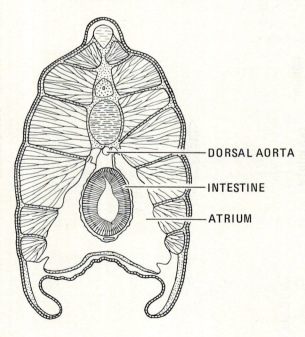

FIGURE 13-5 CROSS SECTION THROUGH INTESTINE

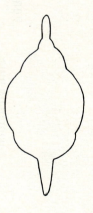

FIGURE 13-6 CROSS SECTION THROUGH TAIL

III. RELATION TO VERTEBRATES

Both the urochordates and the cephalochordates have been proposed as stem forms from which the subphylum Vertebrata arose. A convincing case can be made for the tadpole larva as the progenitor of the vertebrates, in which case the group arose by neoteny. The argument hinges on the idea that the tadpole larva is a new development in the urochordates, as suggested by detailed analysis of the function of the larva in habitat selection mentioned previously. This interpretation assigns the cephalochordates the position of highly specialized filter feeders derived from some intermediate stock in the evolution of the vertebrates.

IV. SUBPHYLUM VERTEBRATA

Various schemes of classification exist for the vertebrates. Our example includes seven extant classes and you should learn to recognize representative specimens of each from the demonstrations in the laboratory.

Class Ostracodermi ("shell skin"). (Extinct.)

Class Cyclostomata ("round mouth"). Lampreys, hagfishes. No jaws, notochord persistent, no ventral fins.

Class Placodermi ("plate skin"). (Extinct.)

Class Chondrichthyes ("cartilage fishes"). Sharks, skates, ratfish. Endoskeleton cartilaginous with some ossification.

Class Osteichthyes ("bony fishes"). Sturgeons, gars, sardines, minnows. Endoskeleton ossified, lung or swimbladder common, two-chambered heart.

Class Amphibia ("double life"). Frogs, toads, salamanders. Usually with limbs, three-chambered heart, moist skin.

Class Reptilia ("creeping"). Snakes, lizards, turtles, alligator. Dry skin with scales, three- or four-chambered heart, columella bone in middle ear.

Class Aves ("bird"). Ostriches, penguins, ducks, canaries, chicken. Feathers and scales, homoiothermic, four-chambered heart, columella and stapes in middle ear.

Class Mammalia ("breast"). Egg-laying echidna, bats, moles, mice, whales, human. Mammary glands, hair, homoiothermic, four-chambered heart, incus, malleus, and stapes bones in middle ear.

QUESTIONS

1. Why do you think the amphibians have received such a name?

2. What do the terms **homoiothermic** and **poikilothermic** mean? **Endothermic** and **ectothermic**?

V. ANIMAL COMMUNICATION: SOUND (VOCAL → AUDITORY) STIMULATION

Interaction among organisms can occur through a variety of sensory pathways that perceive stimuli: tactile, visual, chemical (olfactory), and auditory stimuli. These systems of communication require specific mechanisms of stimulus production and sensory reception. Because our auditory system overlaps the frequency ranges of sound used by many animals for communication, it is possible for us to examine this particular type of biological variation by simply listening to some recorded examples. The purpose of this exercise is to enable you to become familiar with a few of the common animal sounds that you are likely to hear outdoors. To learn them, study the special set of tape recordings that are on reserve in the audio-tutorial office for general zoology students.

A. Amphibian Calls

Anuran amphibians commonly develop male choruses every year that attract females of the same species to breeding ponds. Learn these calls by the **scientific name** of each species.

> *Hyla crucifer* (spring peeper)
> *Hyla versicolor* (grey tree frog)
> *Pseudacris nigrita* (chorus frog)
> *Rana pipiens* (leopard frog)
> *Rana clamitans* (green frog)
> *Bufo americanus* (American toad)

B. Bird Calls

These examples include some specific vocalizations that play important roles in the defense of territory and maintenance of strong pair bonds for breeding birds. Learn these calls by the **common name** of each species.

> Ovenbird
> Chipping sparrow
> Common loon
> Hermit thrush
> Redstart
> Nashville warbler

QUESTIONS

1. Why do you think that it is necessary to learn the amphibian calls according to their scientific name but sufficient to learn the bird calls by their common name?

2. Which of the 12 species has the most "complex" call? The most melodious? The shortest call?

3. Name three other organisms that produce sounds for some type of communication exclusive of birds and amphibians.

4. What is the major mode of communication among humans? Among fishes?

14. FETAL PIG

The fetal, or unborn, pig has been chosen as the mammal for dissection because it has all of the fundamental anatomical features common to the higher mammals, including humans. We might have chosen the cat, the rabbit, or the rat, whose anatomy is very nearly as well known as that of the human, but they are more difficult to dissect or less readily available than the fetal pig.

Keep in mind that the pig which you are studying is an unborn mammal and as such will have characteristics which are limited to the unborn and are lost or modified at birth or shortly thereafter. An obvious structure of this kind is the umbilical cord. There are many other temporary structures, some of which you will observe in the course of dissection.

Besides the study of macroscopic, or gross anatomy, you will study the microscopic anatomy, or **histology**, of the organs and tissues that you dissect. The slides used to study histology are from a wide variety of vertebrate animals. Although they are not identical with the tissues of the pig, they are nevertheless characteristic of vertebrates in general.

As you study each organ system of the pig you should compare your dissection with the demonstration dissections of the dogfish shark and bullfrog that will be available in each laboratory. You must read pertinent sections of your textbook as you prepare for each laboratory period.

I. GENERAL EXTERNAL ANATOMY

Note the primary divisions of the body — **head, neck, trunk, tail,** and the **forelegs** and **hind legs.** Compare these divisions with those of your own body. Probably you have already identified the cord extending from the middle of the underbody as the **umbilical cord,** with which the fetus was connected to the **placenta** in the **uterus** of the mother. Look at the demonstration showing this relationship. At birth, most of the umbilical cord is bitten off by the mother and the rest degenerates. The healing over of the spot where the umbilical cord joined your own body left a scar, the **umbilicus** or **navel.** Now examine the body of the pig more carefully, region by region.

- A. HEAD. Includes the **mouth** and **jaws,** the **nostrils,** and the **snout** with its dorsal elevation used in rooting, the **eyes** whose lids are fused together in the small specimens, and the external ears with the auditory tubes leading to the eardrum.

- B. NECK. So thick as to be nearly obscured. This thickness is caused by the great development of the muscles used in rooting.

- C. THORAX OR CHEST. The part of the body with the **ribs, forelegs,** and **shoulders.** Note the division of the foreleg into the **upper leg, lower leg, wrist, foot,** and **toes,** or **digits,** and their partial correspondence with your own arms. This correspondence becomes more evident if you compare the skeletons of pig and humans. In the pig only two digits of the ordinary five, the third and fourth, are fully developed. The second and fifth are reduced, and the first is completely lost. Feel for the ribs and find the posterior border of the thorax. This bony box encloses the lungs, the heart, and the major blood vessels.

- D. ABDOMEN. The part of the body with a soft underwall, from which the **umbilical cord** extends. Note the two rows of **teats,** or **nipples,** on either side of the cord. Within is the **abdominal cavity,** containing **stomach, intestine, kidneys,** and other **viscera.**

- E. SACRAL REGION. Includes **hind legs** and **pelvic bones,** and the attachment of the pelvic bones to the **vertebral column.** Compare the hind legs with the forelegs and with your own.

Find the **anus,** the posterior opening of the digestive tract just below the tail. In the female pig, the **vulva,** in which openings of the reproductive and urinary tracts occur, is found just below the anus. In the male, the external opening of the **penis** may be found in the ventral midline posterior to the umbilical cord. Two swellings mark the scrotal sacs (**scrotum**) on either side of the midline just below the anus.

F. CAUDAL REGION. The tail is small in the pig. It is even smaller in the human, being reduced to the **coccyx,** a structure that may be seen in the skeleton on display as a short extension of the vertebral column.

II. SKELETAL SYSTEM

The skeleton of the pig is similar to that of other mammals. Skeletons of pig, cat, or human (Figure 14-1) are commonly demonstrated in laboratory. Examine the skeletons or models available and, with the help of your instructor, identify the bones labeled in Figure 14-1.

III. GENERAL INTERNAL ANATOMY

Make a line with a colored pencil along the skin from the tip of the lower jaw to a point 1/2 inch in front of the umbilical cord. Fork the lines around the cord and draw a pair of lines 1/2 inch apart back to the posterior limit of the abdominal wall (Figure 14-2). Ask the instructor to check these lines for accuracy. Using your scalpel, cut through the skin and into the soft underlying tissue along these lines. Now separate the thin layer of skin from the underlying (connective) tissue with the blunt end of your forceps or scalpel for about an inch on one side of your incision. Cut away a piece of skin about an inch square and examine it carefully. It contains many **hair follicles,** which show as bumps on the underside. Are you certain now of what makes the pits in pigskin? Is the skin soft? Tough? Elastic? Any other characteristics?

Cut completely through the body wall just in front of the umbilical cord and, using your scissors, cut anteriorly through the **sternum** (breastbone), being careful not to injure closely underlying structures. Now make the parallel posterior cuts through the body wall. Pour out the liquid contents of the body cavities (it may be stained brown from internal bleeding).

A. Abdominal Region

Refer to Figure 14-3 as you study the structures in the abdominal region.

1. UMBILICAL STRUCTURES

    A white cord will be seen in the abdominal cavity extending anteriorly from the umbilical cord. This is the umbilical vein, which drains the fetal blood from the placenta and into the liver. You may sever this vein, but tie a string around it so you will remember it later. Three cords will be visible beneath the flap of body wall now continuous with the umbilical cord. The lateral pair are the **umbilical arteries,** through which the fetal blood flows to the placenta. The larger hard-walled sac in the center is the **urinary bladder.** The allantoic duct carrying the nitrogenous wastes from the bladder runs through the umbilical cord to the placenta. The fetus thus receives oxygen and food from the mother through the placental circulation and also releases carbon dioxide and nitrogenous wastes there. Make a fresh cut across the umbilical cord and locate in cross section the umbilical arteries and vein and the duct from the bladder.

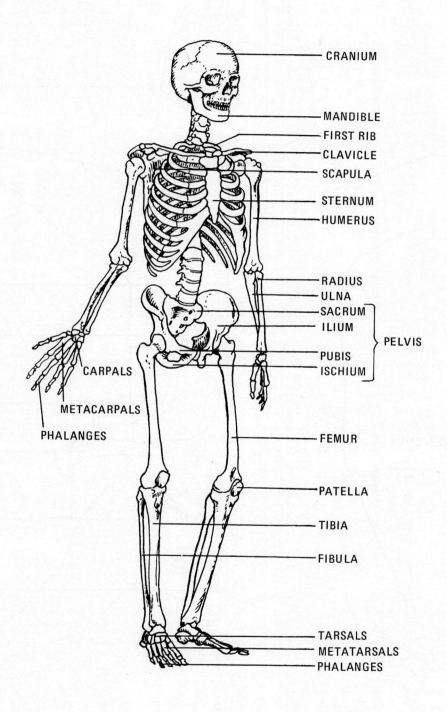

FIGURE 14-1 HUMAN SKELETON, ANTERIOR VIEW

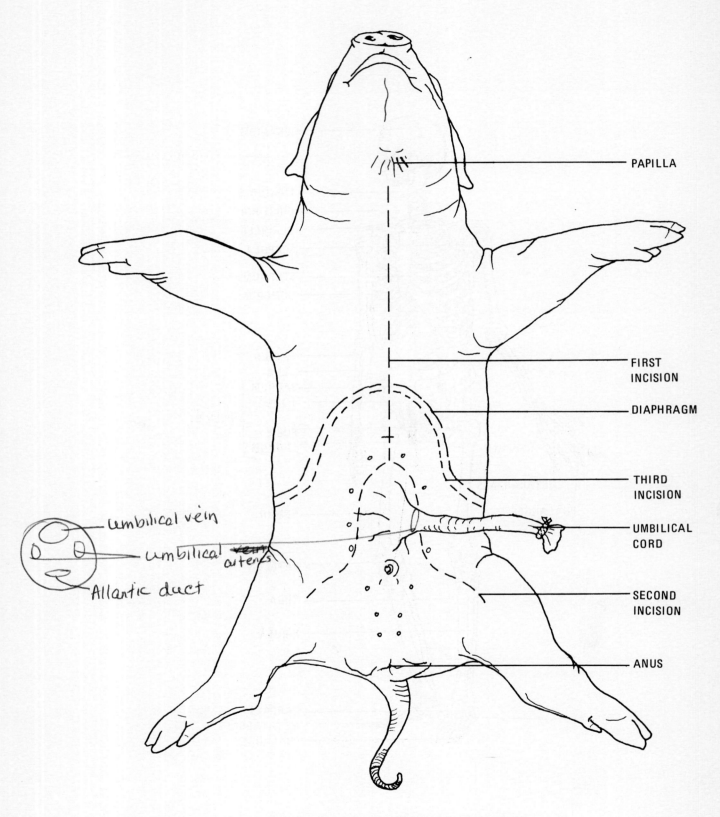

FIGURE 14-2 FETAL PIG, WITH INCISION LINES

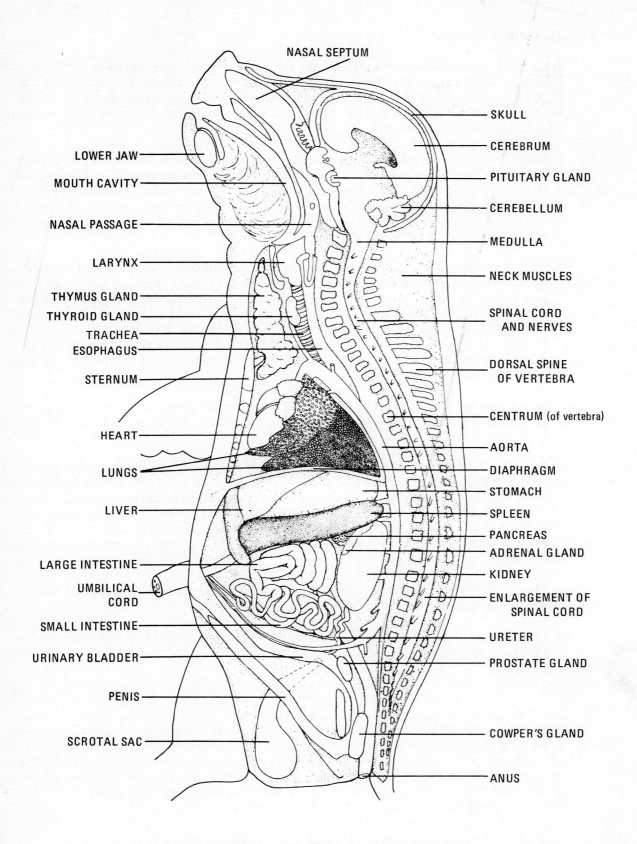

FIGURE 14-3 FETAL PIG, GENERAL INTERNAL ANATOMY

2. ABDOMINAL VISCERA

Make a pair of lateral incisions through the body wall on each side just in front of the hind legs. Wash out the abdominal cavity with cold water to remove any remaining coagulated blood. Examine the viscera (the soft internal organs), separating them with your fingers, being careful not to tear or otherwise mutilate anything. You may refer to Figure 14-4 in addition to Figure 14-3.

   a. Most obvious are the loosely coiled **small intestine** and the thicker tightly coiled **large intestine.** They are supported by a thin sheet of tissue, the **mesentery,** which suspends all of the intestines from the middorsal body wall. This mesentery is formed of two layers of tissue, one continuous with the lining of the abdominal cavity on one side, and the other layer continuous with the lining on the other side. The lining of the abdominal cavity is called the **peritoneum.**

   b. Find the soft, white-walled **stomach** anterior to the intestines, with the dark-colored **spleen** along its left posterior border. A light-colored roughly granular structure, the **pancreas,** is found in the mesentery between the stomach and the first portion of the small intestine, the **duodenum.** Anterior to the stomach is the dark liver. How many lobes of the **liver** are there? Note in one of the liver lobes a dark green sac, the **gall bladder.** It is green because of its contents, the bile. Its duct, called the **bile duct,** drains into the duodenum. The **pancreatic duct,** less easily traced, opens into the duodenum at about the same place.

   c. Now shove the intestines forward and find the posterior continuation of the large intestine, the **rectum.** It is partly concealed by the urinary and reproductive ducts, which will be studied later.

   d. Behind the peritoneal lining of the dorsal part of the abdominal cavity are the relatively large, hard **kidneys.** They are best shown in Figures 14-3 and 14-4. Cut the peritoneum along the lateral border of the left kidney and peel it off toward the midline. Look for the left **adrenal gland,** a crescent-shaped body under the peritoneum immediately in front of and closely applied to the kidney. In life the adrenal glands are orange in color, but in death they are brown because they contain a large amount of coagulated blood.

   e. Finally, explore the **diaphragm,** a membrane, thin in the center but thickened with muscles around the periphery, which separates the abdominal and thoracic cavities.

   f. Examine the demonstration specimens of the dogfish shark and bullfrog and compare the above structures with those of your pig.

B. **Thoracic Region**

1. Cut through the diaphragm on either side where it joins the body wall and pull the **thoracic cavity** open. The thoracic cavity contains the **lungs** and the **heart.** The lungs appear as solid bodies (they contain no air in the fetus). The several lobes of the left lung, the several lobes of the right lung, and the heart are each enclosed in a separate coelom. Each is covered by a layer of peritoneum which is connected by an anterior mesentery to the peritoneum that lines the coelom. Thus, the heart lies within the pericardial sac and between the **pleural sacs,** which in turn contain the lungs.

Cut or break the ribs along each side about two-thirds of the way toward the dorsal midline. Spread the body walls apart, but keep them intact. Examine the **lungs**. How many lobes are there on each side?

2. Notice the two large masses of whitish granular tissue that conceal part of the heart and anterior blood vessels. These are the **thymus** glands, which are comparatively large in young animals but degenerate after sexual maturity. They extend well into the throat region, where they must be distinguished from muscles, particularly those which move the voice box, or **larynx** (Figure 14-3).

    Another gland you will find in the throat region is the relatively small, oval **thyroid gland**, flat against the **trachea** (Figure 14-3) just in front of the thorax. Normally reddish in color, the gland may be cream-colored or brownish in the pickled specimen.

    Also find the **salivary glands** at this time. Pull the skin further away from the left side of the head, up to the eye and behind the ear. Two of the glands, the **submaxillary** and the **sublingual**, are found close together just under the posterior angle of the jaw. These are compact and look a good deal like the thymus. The third salivary gland, called the **parotid**, is very diffuse in structure and less easily found. It is located below and behind the ear, usually plastered tightly to the skin. (It is the parotid gland that is susceptible to the mumps in humans.) Be careful not to destroy the blood vessels and nerves in the neck region.

3. Now return to the examination of the heart. Find a pair of thickened white strands running along each side of the pericardial sac. These are the **phrenic nerves**. Dissect them free of the pericardium, and note their distribution to the diaphragm.

    Remove the pericardial sac from around the heart, being careful not to sever any blood vessels or nerves. Note that this membrane is firmly attached only where the blood vessels enter or leave the heart. The heart obviously consists of several parts (Figure 14-5), the two earlike lappets, or **auricles**, at the front and sides, and the fused **ventricles** in the center and posterior. A diagonal line, marked by branches of a **coronary artery** and a **coronary vein**, passes across the ventral surface and separates the borders of right and left ventricles. Find other branches of the coronary vessels marking the separation of auricles from ventricles. (In what ways are the coronaries medically important to humans?)

4. You should now be prepared to observe the major arteries and veins in their relation to the heart. Recall that a **vein** is a thin-walled vessel which transports blood **toward** the heart, and an **artery** is a thick-walled vessel carrying blood **away from** the heart. The arteries subdivide in the various tissues of the body into capillaries, very fine, extra thin-walled tubules which recombine into veins. The circulation, then, involves passage of the blood in sequence through the heart, arteries, capillaries, and veins, which return the blood to the heart. Watch the circulation of the blood of the frog through the capillaries in the web of the foot.

    a. Anteriorly, several veins from the head and neck, and from the shoulders and forelegs, unite to form the **precaval vein**, which enters the right auricle from the front. These veins are located ventral to the arteries supplying the same regions, but are thin-walled, usually empty of blood, and therefore difficult to follow. Pressing on the auricles with fingers often will drive fluid into them and will enable tracing them if they haven't been cut. Refer to Figure 14-5 and identify as many of these veins as you can in your specimen. You may remove the muscles extending from the sternum to the larynx and head as necessary to expose the veins.

b. Find the **postcaval vein** and follow it from the place where it penetrates the center of the diaphragm, through a groove in the small median lobe of the lung, and into the right auricle from behind. This vein brings back all the blood from the posterior part of the body.

c. Two large arterial trunks are seen to leave the ventricles anteriorly. The most ventral of the two is the **pulmonary trunk,** which delivers blood directly to the lungs. It cannot be traced fully without further dissection, which will be made later. The other trunk, the **aortic arch**, at once gives off two main arteries which send branches into the foreleg and the regions of the shoulder, the neck, and the head.

Lift the left lung so you can follow the aortic arch dorsally. It turns posteriorly and runs along the dorsal midline as the **dorsal aorta,** eventually delivering blood to the entire posterior part of the body.

d. Just ventral to the dorsal aorta is a thick, white tube, the **esophagus,** which leads from the mouth, through the diaphragm, and into the stomach. Along the esophagus run two main branches of the **vagus nerves.** The vagus nerves supply, or innervate, the thoracic and abdominal viscera. Another pair of nerve cords, bearing a series of small swellings along their course, may be seen behind the peritoneum, dorsal and lateral to the aorta on either side. These are the main trunks of the **sympathetic nervous system.**

5. At this point, take inventory of the structures you have observed so far, according to the names underlined in the guide, and answer the following questions:

a. What are the parts of the digestive system seen so far?

b. What other abdominal structures have you identified?

c. What structures occur in the chest cavity?

IV. CIRCULATORY SYSTEM

External respiration is only one phase of gaseous exchange in the animal body. Of primary importance is the other phase, **internal respiration,** involving the absorption of oxygen and giving off of carbon dioxide by all the tissue cells of the body. The circulatory system has an integral role to play in linking the two phases of respiration. Through the medium of the blood, it takes oxygen from the lungs to the body tissues, where internal respiration occurs, and brings carbon dioxide to the lungs, where external respiration takes place. Its function in this respect is thus one of transport. The next question to be answered, then, is as follows: In what steps is the transport of the blood through lungs and body tissues accomplished? To answer this question, a careful and detailed dissection of the heart in its relationship to the lung (**pulmonary**) and body (**systemic**) circulations is necessary.

Trace the major branches of the systemic arteries and veins, including one interposed venous system, the hepatic portal. This will be done only in the detail necessary to understand whence the major organs get their blood supply, what happens to the blood in them, and where the blood goes after leaving them. Arteries are shown in Figure 14-4 and veins in Figure 14-5. It will be well to color the arteries red and the veins blue.

A. **Arterial System**

Find the **innominate** artery, which gives rise to several arteries supplying the chest region, the shoulder, and the foreleg. Trace these as far as they proceed freely.

Very shortly the innominate artery divides into the **right** and **left common carotid** arteries, which carry all the blood that goes to the head region. There they branch into the **external carotids** to the face and the **internal carotids** to the skull cavity and brain. Do not attempt to trace the latter vessels any distance.

Separate a pair of nerve cords from one of the common carotid arteries — all run in a common sheath. The nerves are the main trunks of the **vagus** and **sympathetic** systems. These were seen before, separately, in the thoracic region. Trace the nerves on one side to the point where they separate.

Turn your attention now to the **dorsal aorta**. It gives off a series of small **segmental arteries** dorsally between the ribs, and one or more small arteries ventrally to the esophagus. To trace the aorta into the abdominal cavity, cut directly through the diaphragm (between a pair of heavy muscle bundles) to the aorta. Right at the point where the aorta penetrates the diaphragm, the **coeliac artery** arises and branches to the spleen, pancreas, stomach, and liver. About 1/4 inch posteriorly, the larger **superior mesenteric artery** comes off the aorta, to branch to all parts of the small intestine and the coiled part of the large. These vessels are exceedingly difficult to find and trace because they are literally covered with ganglia and nerves and with the tissues of the mesentery.

Dissect the left kidney free from the dorsal body wall, and turn it to the right to expose more of the dorsal aorta. Find the **renal artery**. Put the kidney back and continue to explore the aorta posteriorly. A pair of fairly large **iliac arteries** pass laterally to the hind legs. Slightly posterior to these find the **umbilical arteries**, which supply the bladder and placenta. After birth the umbilical arteries degenerate into a pair of small vessels supplying only the urinary bladder. The most posterior extension of the aorta, the tiny caudal artery, will be seen later at the time of the dissection of the reproductive system. If your pig is a female, be extremely careful not to damage the ovaries, oviducts, and uterus, which are suspended by mesenteries from the posterior region of the abdominal cavity.

B. **Venous System**

Look just dorsal to the iliac arteries to find the **iliac veins**, which unite to form the **postcaval vein**. As the postcaval leads anteriorly, it receives **segmental veins** from the body, genital veins from the reproductive structures, and **renal veins** from the kidneys. Notice that the postcaval turns to the right around the dorsal aorta and comes to lie almost ventral to it at the level of the renal veins (Figure 14-5).

From this point the postcaval is not so easily traced. It is embedded dorsally in the extreme right lobe of the liver. Turn the intestines and liver to the left, and scrape the liver tissue away from the point where the vein enters the liver to the point where it leaves and penetrates the diaphragm. In the anterior part of the liver, it receives several **hepatic veins** — at least one from each lobe — and also a small connection from the **umbilical vein**. This connection, difficult to find in the hard liver tissue of the preserved fetus, is lost after birth.

The **umbilical vein** distributes itself for the most part to the various lobes of the liver, where it is continuous with parts of the **hepatic portal vein**. (The hepatic portal vein begins in capillaries

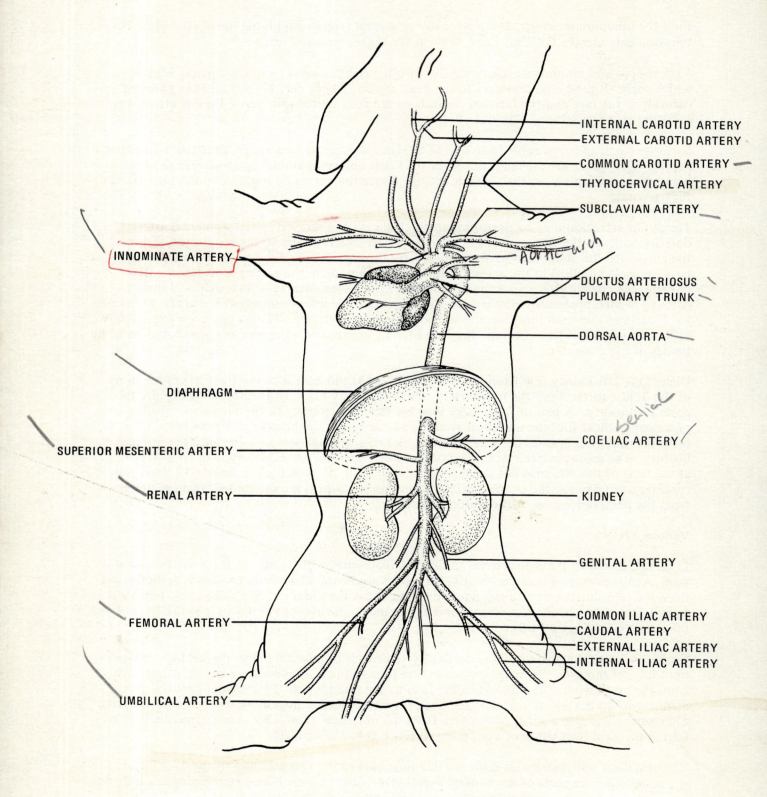

FIGURE 14-4 FETAL PIG, MAJOR ARTERIES

90

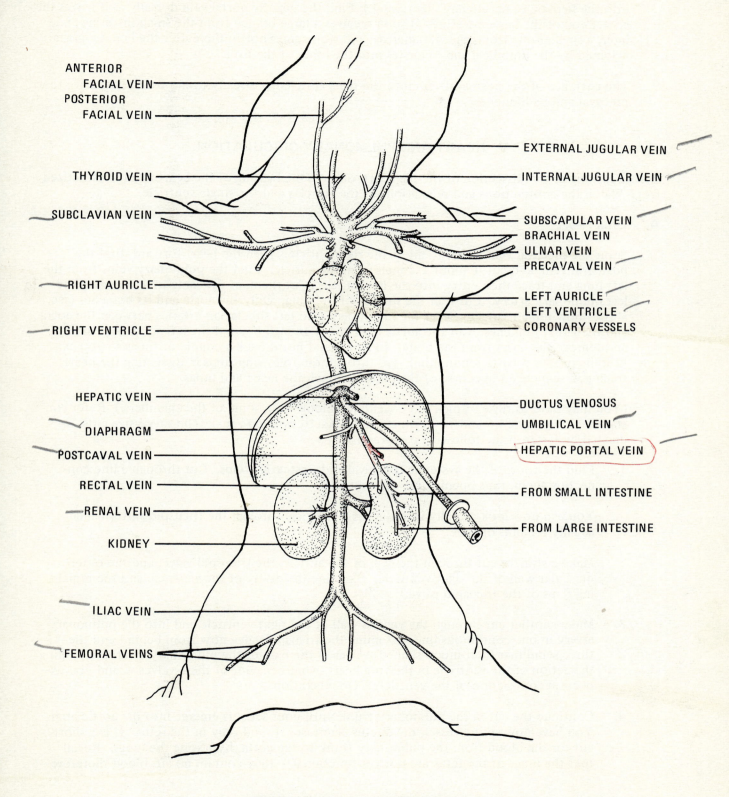

FIGURE 14-5 FETAL PIG, MAJOR VEINS

91

in the small and large intestines and ends in capillaries in the liver. The liver capillaries collect into the hepatic veins, already mentioned.) Find the hepatic portal vein dorsally as it leaves the center coil of the large intestine. It soon receives a large branch from the small intestine and, later, a smaller one from spleen, pancreas, and stomach. Shortly thereafter the hepatic portal is joined by the umbilical and branches into the lobes of the liver.

The entrance of the postcaval vein into the right auricle has already been observed, as have the precaval and its branches.

V. HEART AND PULMONARY CIRCULATION

Before turning to the dissection of the pig heart, observe the beating of the heart in an anesthetized frog. Watch the contractions and relaxations of the auricles and the single ventricle.

A. Fetal Heart

Sever the **precaval** vein where it enters the **right auricle**. Turn the heart forward, find the **postcaval** vein, and cut it about 1/2 inch from the auricle. Find the **pulmonary** veins from the left lung and trace their entry into the left auricle. Cut them, at the same time exposing the **left pulmonary artery**. Cut it at the surface of the lung. Only the aorta and its branches (see above) now remain intact. Sever the innominate and left subclavian arteries between the aorta and their first branches. Cut the **aortic arch** about an inch beyond the left subclavian artery and remove the heart from the body. Examine and put it back a couple of times so you can keep its proper orientation in mind. Go through the following steps in dissecting the heart, tracing in sequence the course of the blood through the heart and lungs.

(You may have to rinse or pick out coagulated blood from some of the chambers.) Refer constantly to Figures 14-6 and 14-7, which show exactly where to make the successive cuts, indicated by letter in the following paragraphs.

1. Find the roots of the **systemic** (precaval and postcaval) **veins**. Cut through a line connecting them, thus opening the **right auricle**.

    You can now look directly into the right ventricle through the **tricuspid valve**, which usually remains open.

2. Make a straight cut through the wall of the auricle, the tricuspid valve, and the outer muscular wall of the right ventricle. Examine the cavity of the ventricle and the remaining flaps of the tricuspid valve.

3. Make another cut through the ventral wall of the right ventricle and into the **pulmonary artery** in one continuous line. Examine the tricuspid valve now from behind and the three **semilunar** (half-moon shaped) valves in the base of the pulmonary artery. Recalling the action of the ventricle in the frog heart, what behavior of these valves would you expect on contraction of the ventricle? On relaxation?

4. Continue the cut in the pulmonary trunk until your scissors emerge into the **aortic arch**. You have now split a vessel, the **ductus arteriosus**, found only in the fetus. It is a short-cut for the blood from the pulmonary trunk to the **aorta**, bypassing the lungs. Recall that the lungs of the fetus are not yet functional — they contain no air; blood therefore

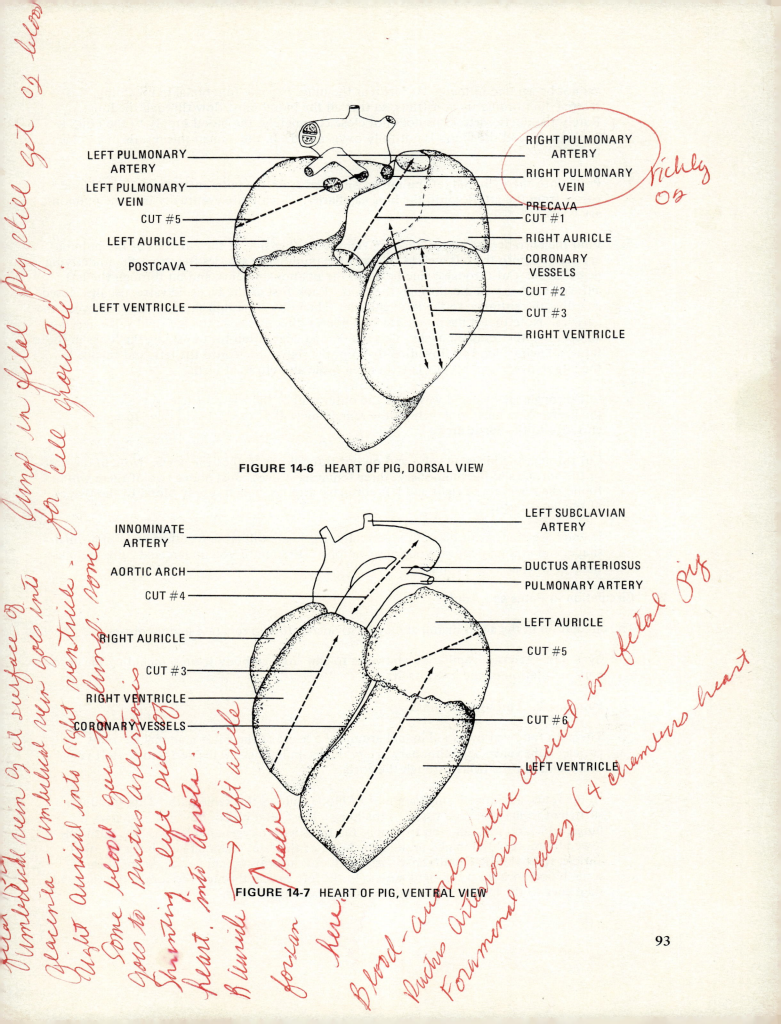

FIGURE 14-6 HEART OF PIG, DORSAL VIEW

FIGURE 14-7 HEART OF PIG, VENTRAL VIEW

cannot be aerated in them. At birth, as the lungs become functional in breathing, this arterial duct ordinarily constricts, so that all the blood must flow through the lungs. Rarely the ductus arteriosus does not close completely, the blood supply to the lungs is inadequate, and the blood is not fully oxygenated. In the human this results in a "blue baby."

Probe into the right and left **pulmonary arteries** from the pulmonary trunk. The blood passes from the arteries into the **lung capillaries**, which unite eventually into the **pulmonary veins**.

5. Find the entrances of the pulmonary veins, on either side of the cut ends of the pulmonary arteries, into the left auricle. Dissect the pulmonary arteries free from the underlying heart tissue. Cut through a line between the pulmonary veins, and on through the left auricle about as you did through the right. Look first at the **bicuspid valve**, usually closed, guarding the entrance to the **left ventricle**. Then observe a thin sheet of tissue between the auricles. Note that it has an unattached border directly above the bicuspid valve, leaving an opening, the **foramen ovale**. This is a second adaptation for bypassing the lungs in the fetus, allowing blood to flow directly from the right auricle into the left, and thence to the left ventricle and aorta. The foramen ovale also closes at birth.

6. Cut through the bicuspid valve and the outer wall of the left ventricle to the very tip of the heart. You must look behind the remaining wall of the bicuspid valve to see the entrance to the **aortic arch**.

7. Cut through the bicuspid valve and on out through the wall of the aorta. Note in the base of the aorta, as in the pulmonary, three **semilunar** valves. Just above two of these will be found the openings of right and left **coronary arteries**, which supply blood to the tissues of the heart itself.

If you have made your dissection correctly, the heart will close together again in normal shape, and you may repeat your observations. Do so until you understand fully the relations of the parts of the heart.

B. Circulation in Fetus and Adult

The circulation in the adult mammal may be divided into two separate major courses:

1. Systemic. Arteries, capillaries, and veins to and from all body tissues, including the coronary circulation of the heart itself.

2. Pulmonary. Through the lungs.

The path followed by blood in and around the heart of the adult mammal is: right auricle, right ventricle, pulmonary artery, lungs, pulmonary veins, left auricle, left ventricle, and out the heart through the aorta.

In the fetus, a relatively small amount of blood follows this same path. This amount is sufficient to maintain the tissue of the lungs. Most fetal blood, however, bypasses the lungs, which do not function to supply the fetus with oxygenated blood. The path followed by most blood in and around the heart of the fetal mammal near birth is: right auricle, right ventricle, pulmonary artery, ductus arteriosus, and aorta. As mentioned, some blood goes from the right auricle to the left auricle through the foramen ovale, and then goes to the left ventricle and the aorta. See Figure 14-8.

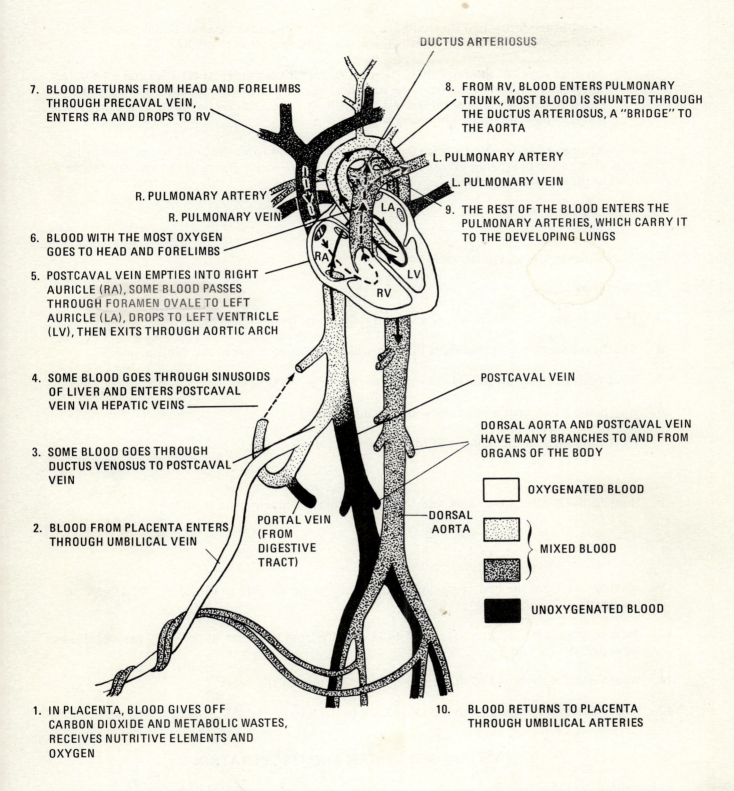

FIGURE 14-8 CIRCULATION IN FETUS

Thus, in the fetus, two shortcuts are available to bypass the nonfunctional pulmonary circulation. What are they? What is the result of their closure at birth?

Another vessel of the fetal circulation is the ductus venosus (Figures 14-5 and 14-8). It carries oxygenated blood from the umbilical vein to the postcaval vein, enabling some blood to bypass the liver on the way to the heart. After birth, the ductus venosus closes and becomes the ligamentum venosum.

You will be expected to be able to trace the course of the blood from the heart to any part of the body and back to the heart and, eventually, to understand the major chemical changes that occur in the blood in the various capillary beds.

QUESTIONS

1. What is the action of the tricuspid valve?

2. The bicuspid?

3. The semilunar valves in the aortic arch and pulmonary trunk?

4. What is the sequence of chambers through which the blood flows in the adult mammal?

5. What parts of the heart provide motive force to the blood?

6. Why is the left ventricle more muscular than the right?

7. What functions do the muscles of the auricles perform?

8. How do the hearts of the shark and frog differ from the mammalian heart?

9. What vessels from the body (systemic veins) carry blood directly to the heart? What part of the heart?

10. What vessels (systemic arteries) carry blood directly from the heart to the body tissues? From what part of the heart?

11. What vessels carry blood to the lungs? From what part of the heart?

12. What vessels carry blood from the lungs back to the heart? To what part of the heart?

VI. RESPIRATORY SYSTEM AND ITS OPERATION

In breathing, air taken into the **nostrils** passes through the **nasal passages** and into the **pharynx** at the back of the mouth cavity. Here the **alimentary** (food) and **respiratory** passages cross. The air goes ventrally into the **larynx**, and the food passes dorsally into the **esophagus**. To see these relationships, dissection is necessary. Refer to Figure 14-3.

1. Insert your scalpel successively into the corners of the mouth and cut directly back to the jawbones on each side. This severs the muscles which hold the jaw closed. Open the mouth and inspect the **oral cavity** and **tongue**.

    Cut completely through the lower jaw and tongue (in the midline, anterior to posterior, ventral to dorsal) so as to separate them into equal halves, left and right. Separate the halves of the jaw (note that the jawbones have not yet fused in front), and examine the posterior region of the pharynx. Find the **epiglottis**, a flap of tissue that is attached to the ventral side of the pharynx and surrounds the **glottis** (the opening of the larynx into the pharynx). The epiglottis is frequently inserted into the posterior openings of the nasal passages. What is the function of the epiglottis?

    Continue the midventral cut into the larynx and **trachea**. Determine the relationship of the larynx to the nasal passage and the esophagus. A flexible probe or wire may be inserted into the posterior opening of the nasal passage on one side and pushed to a point where it may be seen in the nostril on that side. This passage will be dissected later.

2. Expose the trachea back into the thorax as far as the right lung. Slit the trachea a short distance. Note the successive rings of cartilage which reinforce its walls. Are the rings complete dorsally?

3. Cut the **bronchial tubes** (which connect the trachea to the lungs) close to the lung. Now remove the right lung from the chest cavity by severing any remaining mesenteries. Study, at the root of the lung, the cut ends of the pulmonary veins and arteries and bronchial tubes for thickness of wall and diameter of the passages. Notice now, if you haven't already done so, that the small median lobe of the lung, which partially surrounded the postcaval vein, actually is a part of the right lung.

4. Granting that the left lung is supplied in a manner comparable with the right, consider the mechanical factors operating in **external respiration**, or breathing.

    a. The thorax is a closed box. It may be enlarged by:

    (1) contraction of **intercostal muscles** between the ribs. The ribs, directed toward the posterior at rest, move together, forward, and outward when the intercostal muscles contract, thus enlarging the chest capacity. (Expand your own chest and visualize how this process operates.)

    (2) contraction of the **muscles** of the **diaphragm**. At rest, the diaphragm is dome-shaped and bulges into the thoracic cavity as you saw it. On contraction, it flattens, the dome disappears, and the chest cavity increases in volume. This increase in volume results in a decreased air pressure in the lungs, so that air rushes into the lungs from the outside to equalize the pressure, outside and inside. Air is said to be **inhaled**.

    b. **Exhalation** (breathing out) is accomplished mostly by contraction of the **abdominal muscles** (and certain chest muscles other than the intercostals), which force the diaphragm upward and the thoracic cage downward. Swift action of these muscles produces coughing. To **feel** the chest constrict and the abdominal muscles become taut, exhale rapidly.

QUESTIONS

1. What happens to the temperature of the air you take in?

2. What happens to the moisture content of the air you breathe?

3. What happens to oxygen and carbon dioxide in the lungs?

4. What happens to your **respiratory rate** after exercise? What is your present respiratory rate? Measure it.

VII. DIGESTIVE SYSTEM

The digestive and respiratory systems are the two which provide for **intake** of substances into the body. They are correlated in that the processes of **digestion** prepare food for absorption and use, whereas **internal respiration** provides for the oxidation of food with the release of energy. In addition to energy release, our foods also provide for growth and repair.

1. Turn your attention to the lower jaw. Pull away the membrane of the dorsal surface, so as to expose the "milk teeth" still embedded in the jaw. Extract one. Note that the cavity of each tooth is filled with a jelly. (How many teeth are there at this stage? How many permanent teeth in the adult pig? In the human?)

2. Again examine the region of the pharynx for the origin of the **esophagus,** and trace it posteriorly. Cut through the diaphragm to the point where the esophagus joins the **stomach.**

3. Slit the stomach through the outer edge of the **greater curvature** and look inside. The larger part is known as the **cardiac portion** and the smaller end as the **pyloric portion.** Is there any clear line of demarcation between the two parts? Continue the slit through the wall of the esophagus at one end and through the **pyloric valve** and the wall of the duodenum at the other. Note the heavy muscle in the wall of the pyloric valve; this is called the **pyloric sphincter.** What function does it perform? (How prominent is the **cardiac sphincter** at the esophagus? How much of the process of digestion is completed in the stomach?)

4. The first 2 or 3 inches of intestine is the **duodenum.** What relation does it bear to the pancreas? Remove the small intestine, beginning with the duodenum, by cutting the mesentery and gradually unwinding the entire structure. How long is it in your specimen? (It is about 20 feet long in humans.) About one-half of this length is **jejeunum,** in which most digestion is completed. The remainder is **ileum,** in which most of the absorption of food takes place. Cut open a portion of the ileum and examine the lining. Note that the lining forms a series of macroscopic folds. Many small fingerlike projections, barely visible to the naked eye, cover the internal surface. These are called **villi.** The net result of these folds and villi is to tremendously increase the surface area of the small intestine. Why? How is the surface area of the small intestine increased in the shark (consult demonstrations)? In the earthworm? Find the place where the small intestine joins the large. Extending to the left is a blind pouch 1/2 inch to 1 inch in length, the **caecum.** Unravel the **large intestine** and estimate its length. What are its major functions?

5. The **liver** and **pancreas** arise from the developing digestive tract in the early embryo. The **spleen** acts primarily as a storage center for blood and a generator of white corpuscles. The pancreas acts in part as a digestive gland, secreting pancreatic juice into the duodenum, whereas a primary function of the liver is that of food storage. What is the source of food brought to the liver in the fetus? The source of the food after birth? How is it brought to the liver in each case?

VIII. EXCRETORY SYSTEM

You have seen that the respiratory and digestive systems provide for intake of essential substances, and the circulatory system for their distribution. As food and oxygen are used in metabolism, waste products necessarily are produced. These are, in large part, carbon dioxide, water, and nitrogenous wastes. The CO_2 and part of the water escape from the lungs in external respiration, whereas the rest of the water and the nitrogenous wastes are removed by the kidneys and pass out as **urine**. The kidneys, the urinary bladder, and their associated ducts are called the **excretory system**, and the processes of waste removal **excretion**. For these structures in the female, refer to Figure 14-9, and in the male, Figure 14-10.

Recall that the blood supply to each **kidney** consists of a small artery and a pair of veins connected by a host of capillaries. Obviously, only a relatively small portion of the total supply of blood in the body can pass through the kidneys in any limited period of time. Consequently, the kidneys are able to remove only a small part of the total nitrogenous wastes present in the blood. They act as bailers, continually removing wastes from the blood so as to keep their concentrations within tolerable limits.

1. Remove the left kidney and cut it lengthwise into equal halves. The center is occupied by blood vessels and a relatively large **reservoir,** in which the urine first collects. The layer, which appears in gross section to have many fine striations perpendicular to the surface, contains myriads of excretory tubules, which lead to the reservoir from tiny knots of capillaries called **glomeruli.** Study the demonstrations of the kidney showing glomeruli and ducts.

    The reservoir empties into the **urinary bladder** through the **ureter.** The ureter runs posteriorly to the urinary bladder. Expose the duct but do not injure any discrete structures that cross over it. These structures include the **umbilical arteries** and the **reproductive ducts** in both sexes, and the free-swinging **ovaries** just behind the kidneys in the female. The bladder in turn empties to the outside through another duct, the **urethra,** which will be observed along with the dissection of the reproductive system. During fetal life, however, wastes pass out the **allantoic duct** to the placenta.

IX. REPRODUCTIVE SYSTEM

Normally, metabolism in the young animal leads to growth, maturation and, eventually, to reproduction. In higher animals, two sexes are involved, in which the reproductive organs have essentially similar origin, but finally different structure and function. Because the female reproductive system is less changed from the basic pattern than that of the male, it will be considered first. (Every student will be responsible for knowing the systems in both sexes. Comparison of dissected males and females is therefore necessary.) In both the male and the female, the reproductive system consists of the sex glands, or **gonads,** their ducts, and certain associated glands (Figures 14-9 and 14-10).

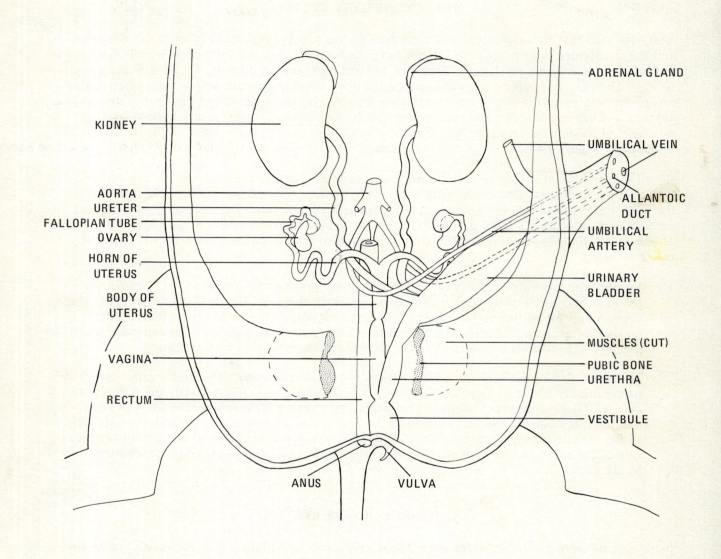

FIGURE 14-9 FEMALE FETAL PIG, EXCRETORY AND REPRODUCTIVE SYSTEMS

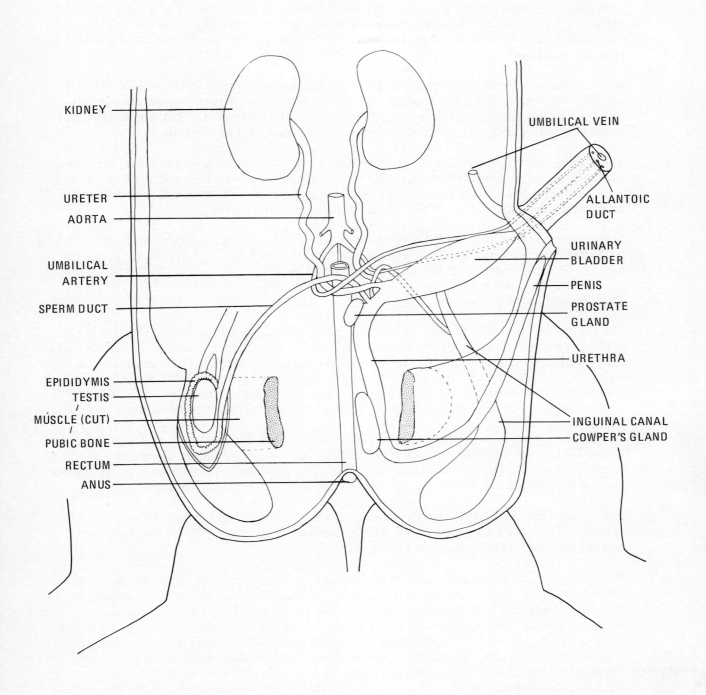

FIGURE 14-10 MALE FETAL PIG, EXCRETORY AND REPRODUCTIVE SYSTEMS

A. **Female Reproductive System**

The **female gonads,** or **ovaries,** are found behind the kidneys swinging rather freely from a pair of thin mesenteries which traverse the kidneys. A tube, very fine in the fetus, is coiled lateral to each ovary. This part of the oviduct is called the **fallopian tube.** The **oviduct** continues on each side as a slightly larger tube, one of a pair of **horns** of the **uterus.** The young develop in the uterus, which becomes vastly enlarged at maturity and pregnancy. The two horns of the uterus unite into the shorter single **body.** In the human, the fallopian tubes enter a single (unpaired) uterus.

Now cut dorsally from the ventral midline until the legs may be separated widely. The cut will pass between muscles and through or between the pelvic "bones," which are actually cartilages at this stage of development. Three tubes will be exposed. They are, in sequence from ventral to dorsal, the **urethra,** the **reproductive tract,** and the **rectum.**

The urethra and the reproductive tract unite about 1/2 inch internal to the **vulva,** forming a common chamber, the **vestibule,** and the tube connecting the vestibule and the body of the uterus is the **vagina.** Open the vestibule and vagina along one side to see the openings of the urethra and the constriction of the vagina at the base of the uterus.

Fully expose the rectum and slit it open from the anus anteriorly. Can you distinguish an **anal sphincter** muscle? Find the tiny **caudal artery** against the spinal column directly above the rectum. It is the posterior extension of what vessel?

B. **Male Reproductive System**

The position of the **male gonads,** or **testes,** depends on the age of the fetus. The testes begin development in the same location as that occupied by the ovaries but gradually descend into the **scrotal sacs** before birth. They may be found anywhere along this path in pigs of the sizes ordinarily dissected. In all specimens the **sperm ducts** may be seen looping over the umbilical arteries and the ureters, to unite dorsal to the urinary bladder. Externally, they penetrate the abdominal wall a little to each side of the midline by way of the **inguinal canals.**

Cut through the skin of one of the scrotal sacs, and extend the incision to the point where the sperm duct leaves the abdominal cavity. The inguinal canal thus will be opened and the **testis** certainly exposed. Open the sac surrounding it. The much coiled tubule looping around the testis is the **epididymis,** which empties posteriorly into the single **sperm duct.**

Cut through the skin slightly to one side of the ventral midline and expose the slender **penis** extending from the anal region to the **urinary orifice.** The central tube of the penis is the **urethra,** the duct which drains the urinary bladder.

To expose the remaining structures, it is necessary to cut through the tissues in the midplane between the legs. Spread the legs wide apart as the incision is completed as far as the **urethra.** The urethra appears to be heavier in the male than in the female due to the presence of certain accessory glands. **Cowper's glands,** about 1/2 inch long and one-third as thick, occur laterally and posteriorly, and the **prostate gland,** oval and about 1/4 inch wide, is found dorsal to the junction of the bladder and urethra. Dissect these structures free of the **rectum,** and expose the point of entrance of the sperm ducts into the urethra at the level of the prostate gland. In the male, the urethra serves for the transport of sperm cells as well as of urine to the outside.

Dissect the rectum free of the dorsal body wall and slit it from the anus forward. Do you find an **anal sphincter** muscle? Observe the tiny caudal artery dorsally against the vertebral column. What is its origin? Its destination?

C. **Operation of the Reproductive System**

In the mature male, **sperm cells** (spermatozoa) are produced in **tubules** in the **testes**. Study the demonstrations of developing sperm cells in a section of mammalian testis and of free spermatozoa of pig and humans. As they are produced, sperm cells are discharged into the **epididymis,** where they remain for an extended period ripening, after which they are capable of fertilizing an ovum. They are discharged in a **seminal fluid,** or semen, which is composed mostly of secretions from the prostate and Cowper's glands.

After the seminal fluid is deposited in the **vagina** of the female, some of the spermatozoa eventually reach the farthest point in the fallopian tube (upper end of mammalian oviduct), where they are ready to unite with any discharged ova (egg cells).

Ova develop in the ovary in little hollow structures called **follicles,** beneath the outer wall. Study the demonstrations of ripening ova in sections of ovaries of cat and human. When ripe, the ova burst out of the follicles and into the fallopian tubes, where **fertilization** (union of sperm and ovum) normally occurs. After fertilization, development of the embryo follows in a complex series of events, including the formation of the placenta in the uterus of the mother, where the greater part of embryonic development occurs.

X. NERVOUS SYSTEM

Coordination of activities in the mammal is accomplished through two general mechanisms: **nervous control** and **hormonal control.** Although their functions may be separated on the basis of mode of action, they are nevertheless interdependent and supplementary. Nervous control depends on direct stimulation of **effector organs** (muscles and glands) by nerve cells. Hormonal control depends on the production of an **excitatory** or **inhibitory chemical** (hormone) in one organ, its distribution by the blood to all tissues, and finally its selective action on a particular organ or tissue different from that which produced it. In brief, a hormone is a substance produced by one tissue, which either stimulates or depresses the activity of another tissue elsewhere in the body.

The nervous system can be divided on the basis of function into two distinct parts. The **voluntary nervous system,** under the conscious control of the individual, consists of the neurons of the thinking apparatus, plus those which receive stimuli from the major sense organs, and those which stimulate the effector organs, the **striated,** or **voluntary, muscles.** The second part is the **autonomic nervous system,** not under voluntary control. The autonomic system governs the so-called vegetative activities of the individual — digestion, secretion, excretion, circulation, and certain others. The effector organs are **cardiac muscle, smooth muscle,** and **glands** — that is, all the effectors **except voluntary muscle.** These effector organs are, in addition, subject to hormonal control.

The **autonomic nervous system** is subdivided into two systems, the **sympathetic** and the **parasympathetic,** both of which innervate the internal organs, and whose actions oppose one another. In general, the sympathetic nervous system stimulates heart action and depresses the activity of the digestive system and associated organs. Conversely, the parasympathetic (which includes the vagus nerves) tends to slow heart action and promote digestive activity. Many additional functions are known, some of which contribute to the emotions of the individual (through the many connections with the central nervous system) without the person being aware of the reason for or the source of his or her feelings. Likewise, some of the activities of the autonomic system are actuated through the voluntary nervous system, for example, salivation (involuntary) at the sight, the smell, or the anticipation of food.

A. **Summary of Relations of the Coordinating System**

1. The autonomic system is composed of sympathetic and parasympathetic systems, which supply a double and opposing innervation to the internal organs.

2. The effectors of the autonomic system are the involuntary muscles and the glands.

3. Hormones affect, among others, structures also under autonomic control.

4. Connections that occur between autonomic and voluntary nervous systems allow the modification of conscious behavior through emotions seated in the autonomic system.

5. The voluntary nervous system receives the stimuli of which we are aware, correlates them, and evokes response to them directly through the action of voluntary muscles, or indirectly through the autonomic system.

B. **Dissection of the Brain**

A complete dissection of the brain and the 12 cranial nerves will not be made, but enough will be done to get some idea of the major features of the brain. First remove the rest of the skin from the head and neck of your pig and cut away the external ears. Remove whatever remains of the upper and lower lips. Turn the pig on its side, left side up. Refer to Figure 14-3 for orientation of the various parts during dissection.

To expose the brain, the roof and side of the skull must be removed. Heavy scissors are best for this operation. Begin at the top of the skull and pick the bones away a small piece at a time. To keep the brain undamaged, free the tough covering of the brain from each bit of bone before it is broken away. The brain coverings are called **meninges**. The tough outer layer is the **dura mater**; the finer inner layer closely following the contours of the brain is the **pia mater**. (An intermediate layer will not be seen.) Several bones will be removed at this point, and now is a good time to identify some of them. Immediately between the orbits is a pair of **frontal bones** (single in humans). Behind them are the **parietals**. Under the **frontals** and **parietals** is the **cerebrum**, consisting of two **cerebral hemispheres**, the thinking part of the brain — the part, that is, with which you cerebrate. Anterior to the frontals are the **nasal** bones, which cover the olfactory organs. Parts of the **sphenoids** will be picked away in the orbital region, plus the **temporal** bone, behind the orbit. The occipital bones form the back of the skull. To remove the latter, the heavy neck muscles must be scraped free and the entire dorsal part of the skull exposed. The occipital bones cover the **cerebellum**, a part of the brain concerned with muscular coordination, and also the **medulla**, the part from which most of the cranial nerves arise.

When the olfactory organ is exposed, make a vertical cut in it. Notice the many folds of tissue, all of which are covered with sensory epithelium. (What might you infer about the sense of smell in the pig on the basis of the extensive development of the olfactory organ?) You may also cut into the nostril from above and follow the nasal passage posteriorly both to the olfactory organ and to the pharynx. Is it a straight tube or does it branch to the olfactory chamber?

After the cerebrum, cerebellum, and medulla are exposed on the one side, split the brain into equal halves, left and right, and remove the left half. Now the brain stem (including the medulla) may be observed in its relation to the cerebrum and cerebellum. The **pituitary gland** may also be found ventrally just behind the optic nerves, in a little pit in the floor of the skull. When you see the location of this gland, you will understand why it is practically impossible to get at it surgically without doing serious damage to nearby structures.

Compare the parts of the brain of the pig with those of the brain of the human (Figure 14-11).

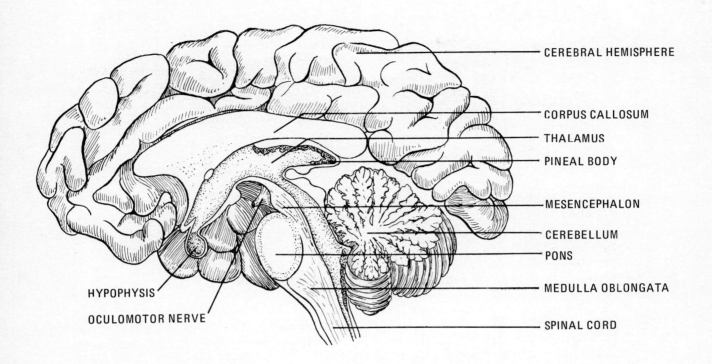

FIGURE 14-11 HUMAN BRAIN, MIDSAGITTAL SECTION

XI. CELLS AND TISSUES

In studying anatomy, it should be remembered that the structures of the body are composed of **tissues** made of microscopic **cells**. You will now examine cells and tissues from some vertebrate animals, representing parts of the body that you have dissected in the fetal pig.

A. **Skin**

Examine the sections of salamander skin (slides labeled **stratified epithelium**) and the demonstrations of human skin. What are some of the functions of the skin or integument in your own body? Refer to Figures 14-12 and 14-13.

B. **Mesentery**

Examine the slide (labeled **pavement epithelium**) of a small section of mesentery showing the cellular details of the two layers. Between these layers occur a large number of blood vessels, which carry blood to and from the intestines, and numerous **lymph glands** (not visible in the slides provided). See Figure 14-14.

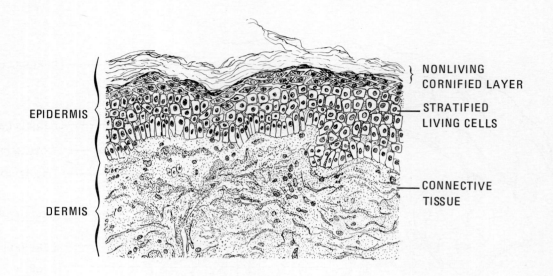

FIGURE 14-12 SECTION OF HUMAN SKIN

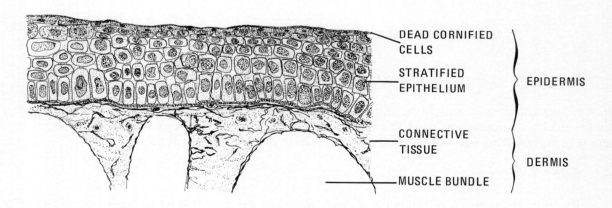

FIGURE 14-13 SECTION OF SALAMANDER SKIN

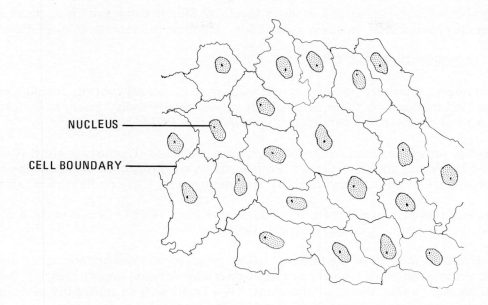

FIGURE 14-14 SECTION OF MESENTERY, UPPER CELL LAYER

C. Small Intestine

Examine sections of the small intestine of frog and cat (slide demonstration labeled **columnar epithelium**), and note the nature of the lining membrane, the presence of capillaries, and the structure of the outer wall. See Figure 14-15.

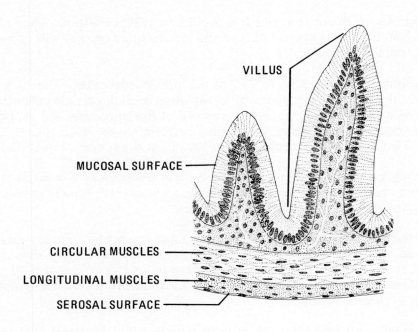

FIGURE 14-15 FROG, SMALL INTESTINE, CROSS SECTION

107

D. **Tissues of Muscular and Circulatory Systems**

Tissues are formed from similar cells grouped together. Muscle tissue and blood tissue are two tissues that are prominent in the mammalian body. Examine the various slides of these tissues.

1. MUSCLE TISSUE, OF THREE TYPES:

    a. **Voluntary**, or **striated**, skeletal muscle, subject to conscious control. Includes muscles used in breathing and those used in body movement generally. Examine both longitudinal and cross sections for details shown in Figure 14-16.

    b. **Cardiac muscle**, striated, but involuntary (not subject to conscious control) (Figure 14-17). Found only in the heart. Propels the blood through the vessels of the circulatory system. Compare, and contrast, its structure with that of voluntary muscle.

    c. **Smooth muscle**, nonstriated and involuntary (Figure 14-18). Occurs in the walls of blood vessels, the digestive tract, and elsewhere.

    The smooth muscles of the urinary bladder are arranged in an interesting way. Examine the demonstration of a small piece of bladder wall showing smooth muscles. Note that the muscle bundles run in all directions. How might such an apparently random arrangement function advantageously in emptying the bladder?

2. BLOOD, THE FLUID CONTENTS OF THE CIRCULATORY SYSTEM

    Examine the slide of human blood and identify the cell types shown in Figure 14-19.

    a. **Red corpuscles**, or **erythrocytes**, are the most numerous cell type in the blood. They are small disc-shaped cells without nuclei. They contain hemoglobin, the red-pigmented substance which acts as a carrier of oxygen and carbon dioxide for respiratory exchange.

    b. **White corpuscles**, or **leucocytes**, which lack color in living condition. They contain nuclei which stain varying degrees of blue and are classified on the basis of nuclear form and granular inclusions.

    (1) **Neutrophils** constitute 65% to 75% of the human leucocytes. The nucleus is lobed (mostly three-lobed), and the small cytoplasmic granules stain a pale neutral shade. Neutrophils are the most actively amoeboid of the leucocytes and readily move out of the vessels into the tissues where they perform phagocytosis (phag = eat, cyt = cell). They form a large proportion of the pus at the site of an infection.

    (2) **Eosinophils** constitute 2% to 5% of the human leucocytes. The granules are large, red, and obscure the usually bilobed nucleus. These cells usually show an abnormal increase in numbers in allergic conditions.

    (3) **Lymphocytes** constitute 20% to 25% of human leucocytes. Three types of lymphocytes are recognized: large, medium-sized, and small. The small, shown in Figure 14-19, is most easily identified. The nucleus stains a deep blue around which the cytoplasm appears as a narrow, blue rim. An abnormal increase in lymphocytes occurs in chronic infections.

    (4) **Monocytes** constitute 3% to 8% of human leucocytes. The large nucleus is pale and stringy and indented in old cells. The abundant cytoplasm is slate gray or muddy blue. An abnormal monocyte number is found in cases of malignancy and infection.

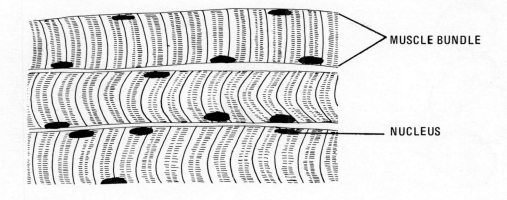

FIGURE 14-16 VOLUNTARY MUSCLE

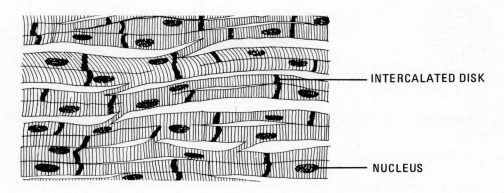

FIGURE 14-17 CARDIAC MUSCLE

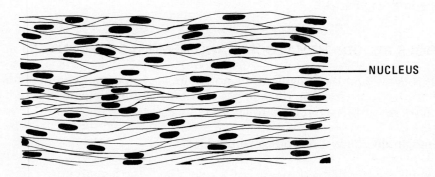

FIGURE 14-18 SMOOTH MUSCLE

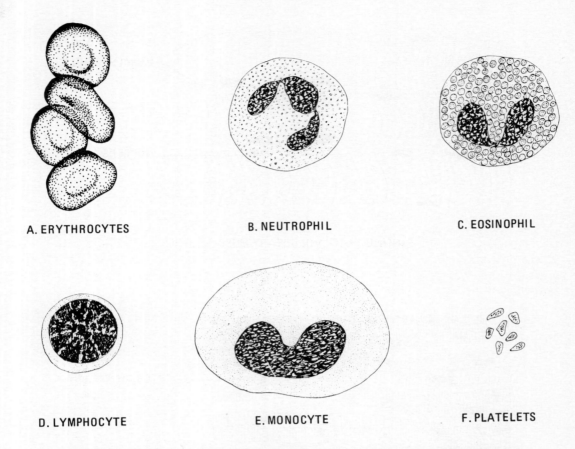

FIGURE 14-19 LEUCOCYTES, ERYTHROCYTES, AND PLATELETS OF HUMAN BLOOD

c. **Blood platelets** are tiny cell fragments without nuclei. They are believed to produce an enzyme which initiates blood clotting.

3. LUNG (DEMONSTRATION) IN SECTION

This tissue is difficult to prepare and study. Note the **air sacs** and the walls of capillaries between them. How much, or how little, tissue separates the air in the lung from the blood in the lung capillaries?

XII. WHOLE BLOOD: HEMATOCRIT AND HEMOGLOBIN CONTENT

Red blood cells, erythrocytes, and their pigment, hemoglobin, are important components of vertebrate whole blood that carry out vital functions of gas transport and participate in acid-base balance of body fluids. The purpose of this exercise is to let you:

1. estimate the **hemoglobin** content of your own blood.

2. estimate the volume percent of erythrocytes of whole blood, the **hematocrit**.

3. develop understanding of a kind of instrument that is commonly used in many types of chemical analyses, the **colorimeter-spectrophotometer.**

4. appreciate variation of biological characteristics (genetic, behavioral, morphological, and physiological) that is present within any species of organism, exemplified here by variation in hemoglobin and hematocrit.

Before attempting any kind of analysis by instrument, it is important to understand the principles on which the technique is based. The colorimeter is composed of four basic elements: light source, monochromator, light-sensitive phototube, and galvanometer. The light source provides a constant input of "white light" energy. The monochromator selects a specific color (wavelength, expressed as nanometres, nm) of the white light spectrum that is to be passed into the sample tube of diluted blood for this exercise. The phototube produces a change in electrical voltage that is proportional to the quantity of light energy that has passed through the sample and onto its light-sensitive surface. And finally, the galvanometer converts these variations of voltage into the mechanical deflection of a needle to provide an output on the scale of our measurement system. The amount of monochromatic light absorbed by a sample of constant thickness is directly proportional to the concentration of absorbing molecules, for example, hemoglobin. The thickness of the sample solution is held constant by use of standardized test tubes. Thus variations in output readings of the galvanometer needle on the **absorbance** scale are directly related to variations in hemoglobin concentration.

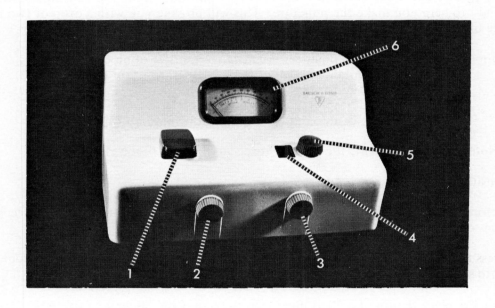

FIGURE 14-20 BAUSCH & LOMB SPECTRONIC 20 COLORIMETER

1. SAMPLE TUBE HOLDER WITH COVER
2. ON-OFF AND PHOTOTUBE CONTROL KNOB
3. LIGHT CONTROL KNOB
4. WAVELENGTH SCALE, NM
5. WAVELENGTH SELECTION KNOB
6. METER, ABSORBANCE AND TRANSMITTANCE SCALES

1. OPERATING PROCEDURES FOR B & L SPECTRONIC 20 COLORIMETER

   a. Turn machine on by clockwise rotation of control knob (2).

b. Select **540** nm on the wavelength scale (4) by rotation of the wavelength control knob (5).

c. Adjust the meter needle to read at "∞" on the absorbance scale (6) by rotation of the control knob (2) when there is **no tube** in the sample holder (1).

d. Place a standardized test tube containing about 5 ml of distilled water in the sample holder (1). For best results, use B & L selected tubes and be sure to align the index mark on the test tube opposite the index line on the sample holder. Be sure that the cover is closed on the sample holder.

e. Rotate the light control knob (3) until the meter needle reads "0" on the absorbance scale with the blank tube in the sample holder.

f. The colorimeter is now ready to read the absorbance at 540 nm of light on an unknown sample, that is, the tube of diluted hemoglobin. For the final step, replace the blank (distilled water) tube with the unknown tube containing 5 ml of dilute blood solution and read the **absorbance** value from the meter.

PROCEDURES FOR SAMPLING BLOOD

Small quantities of blood will be sampled by pricking your clean fingertip with a sterile lancelet. If you have difficulty with this mild form of self-mutilation, ask your instructor or a "friend" to help you make the puncture. This will provide you with a drop or two of fresh, whole blood that can be analyzed for hemoglobin content and hematocrit.

Supplies: 1 20-μl capillary pipet for quantitative transfer of blood
1 test tube containing 5 ml of 0.4% NH_4OH solution
1 sterile lancelet for skin puncture
1 heparinized* capillary tube for hematocrit
1 tray sealing compound to plug tips of hematocrit capillary

*Note: Heparin prevents the clotting reaction of whole blood exposed to air.

3. TECHNIQUE

a. Before letting blood from your fingertip, make sure that you have all of the necessary supplies close at hand.

b. Press the lancelet quickly into cleaned fingertip. Squeeze gently to expel a droplet of blood onto skin surface.

c. Quickly place the tip of a 20-μl pipet into the blood and allow it to fill to the scribed black line. This should occur by capillary action, especially if the tube is held nearly horizontal.

d. Transfer the pipet with 20 μl of blood to the test tube containing 5.0 ml of 0.4% NH_4OH solution. Expel the contents of the pipet by gentle blowing pressure on the attached mouthpiece tubing. Rinse the pipet by repeated filling and draining into the NH_4OH solution. Mix the contents of the tube, and wait for about 10 minutes before reading the absorbance at 540 nm in the Spectronic 20 colorimeter.

e. Squeeze your fingertip again to expel another droplet of blood.

f. Place the heparinized capillary tube into the droplet of blood and allow it to fill to about three-fourths of its total length. This should occur by capillary action, especially if the tube is held nearly horizontal.

g. The ends of the capillary are sealed by pressing them vertically into a tray of sealing compound that plugs the tips. Your instructor will help you arrange the tube in the hematocrit centrifuge.

4. RESULTS

   a. **Hemoglobin.** Record the absorbance of your diluted blood sample:

   Absorbance at 540 nm = _____

   Because hemoglobin concentration is directly proportional to this absorbance value, it can be used directly for comparison with that of other members of your class. However, by knowing the absorbance of a **standard hemoglobin solution** it is possible to convert this value into a direct measure of hemoglobin concentration, that is, grams (g) of hemoglobin per 100 ml of solution, equivalent to grams percent. A standard solution of hemoglobin, **30 grams per 100 ml**, was treated exactly the same as your whole blood, that is, 20 μl diluted into 5 ml of 0.4% NH_4OH and read in the Spectronic 20 against a distilled water blank at 540 nm. The absorbance of **this standard was 1.04**. Now it is simple to convert the absorbance value of your blood into units of hemoglobin concentration, grams of hemoglobin per 100 millilitres of whole blood.

   $$\frac{\text{(absorbance of standard)}}{\text{(concentration of standard)}} = \frac{\text{(absorbance of sample)}}{\text{(concentration of sample)}}$$

   Therefore:

   $$\text{(sample concentration)} = \frac{\text{(sample absorbance)}}{\text{(standard absorbance)}} \times \text{(standard concentration)}$$

   For example, if your sample of diluted whole blood had an absorbance of 0.52 against a distilled water blank at 540 nm, the estimate of hemoglobin concentration would be:

   $$\frac{(0.52)}{(1.04)} \times (30.0 \text{ g}/100 \text{ ml}) = 15 \text{ g hemoglobin}/100 \text{ ml blood}$$

   Convert the observed absorbance of your diluted blood into units of hemoglobin concentration: _____ grams per 100 ml of blood.

   b. **Hematocrit.** The percent, by volume, of packed red blood cells is easily estimated after centrifugation of whole blood contained in heparinized capillary tubes. After the hematocrit capillary tubes have been spun for 5 minutes at 8,500 rpm to pack the cells, remove the tube and measure the length occupied by red cells and the length of tubing taken by cells and plasma. The ratio of these two lengths gives the proportion by volume of packed erythrocytes. These measurements can be most readily done with the hematocrit readers that are available in each laboratory. Hematocrit of your whole blood: _____ %.

5. ANALYSIS

   a. The mean value of hemoglobin concentration.

   If we add all the sample observations of your laboratory class and divide the sum by the number of observations, we obtain the **average**, or arithmetic mean. This sample mean, $\bar{x}$, can be thought of as an expected value for the sample, and can be used to compare one sample with another.

   $$\text{mean} = \bar{x} = \sum_{i=1}^{n} x_i/n = \frac{x_1}{n} + \frac{x_2}{n} + \ldots + \frac{x_n}{n}$$

   b. The range of values of hemoglobin concentration.

   One measure of variability of observations about the sample mean is the difference between the extreme values, the **range**. If we subtract the lowest observed value from the highest value, we obtain the range.

   $$\text{range} = (x_{hi} - x_{lo})$$

   c. Calculate the **mean** and **range** of hemoglobin concentration for values obtained by all members of your laboratory class.

   class mean:_____ class range:_____

   Now subdivide the estimates of hemoglobin concentration into two groups according to sex, and calculate the mean and range for each group.

   class male mean:_____ class male range:_____

   class female mean:_____ class female range:_____

   Do the data of your class demonstrate that there is a difference in hemoglobin content between male and female blood samples?

6. CORRELATION BETWEEN HEMOGLOBIN CONCENTRATION AND HEMATOCRIT

   After looking at the variability of hemoglobin concentrations, we may ask a simple question about the biological nature of the observed differences. Although sampling errors and instrument variations can contribute to the overall observed variability, we can still ask if, for example, low values of hemoglobin concentration are due to:

   (1) fewer red blood cells, or
   (2) lower hemoglobin concentration per cell, or
   (3) a combination of these two obvious possibilities,

   and **vice versa** for high hemoglobin values.

If the hemoglobin concentration per red cell were constant, we would expect to see a positive correlation between hematocrit and hemoglobin; for example, whole-blood samples with high hematocrit would have above-average hemoglobin concentration. One easy method of correlation is the scatter plot whereby paired values (hematocrit and hemoglobin) are plotted on the coordinates of a graph. We will use the scatter plot technique to look for the presence of a correlation between these two variables. Plot the paired values obtained by each member of your class in Figure 14-21. John Doe (Hematocrit = 44, Hemoglobin = 15) is plotted (●).

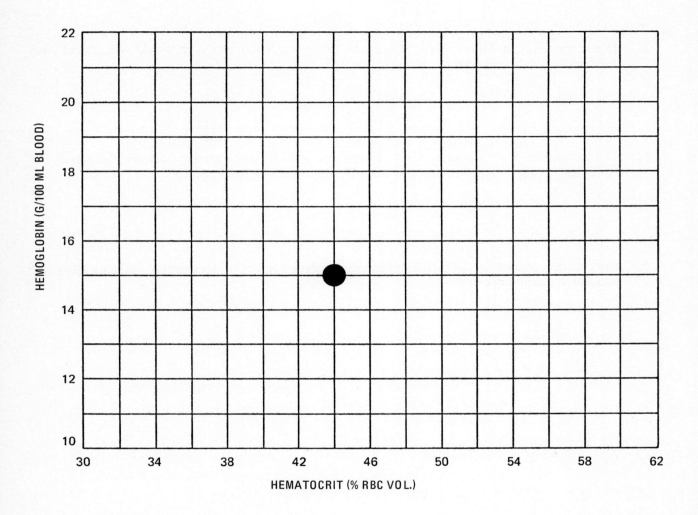

FIGURE 14-21 VALUES OF HEMATOCRITS AND HEMOGLOBIN

QUESTIONS

1. What color is light of 400 nm wavelength? Of 700 nm wavelength?

2. By how many times was your whole blood diluted when 20 μl of it were added to 5.0 ml of NH_4OH solution?

3. If you moved from Minneapolis, Minnesota, to Denver, Colorado, how would you expect your hemoglobin or hematocrit to change? Why?

4. Did the scatter plot of hematocrit $\times$ hemoglobin indicate a correlation between these two variables? What conclusion can you draw by inspection of these results?

5. How could you reduce the variability among observed values of hemoglobin concentration that was due to sampling error?

15. TAXONOMY OF MAMMALS

Most of us, as animals, have special interest in the class Mammalia to which we belong in both an evolutionary and taxonomic sense. This exercise will acquaint you with some of the variation (specialization) that is present in the skulls of modern mammals. Adaptive radiation of mammals has resulted in marked variation in size and shape of skulls and numbers of different types of teeth. The variation in cranial features is used in the taxonomic treatment (the classification and identification) of mammals.

Skulls have been collected from common, local species of mammals and cleaned for your examination and identification to proper **order**. The following artificial key is designed for identification of skulls from adult specimens of five common orders. Choose a skull and examine its form and kinds and numbers of teeth with the aid of Figure 15-1.

Now use the artificial key to identify the order to which the skull of the mammal you have examined belongs.

Artificial key to five common orders of mammals based upon skull characteristics:

1. A vacant space or gap (diastema) between teeth in front of mouth (incisors) and rest of jaw teeth (premolars and molars); no canine teeth 4

   No conspicuous diastema between incisors and rest of jaw teeth; canine teeth present. 2

2. Canine teeth smaller than teeth preceding them . . . . Insectivora

   Canine teeth larger than teeth preceding them. 3

3. Small size, length always less than 25 mm; gap between upper incisors; jaw teeth with high, pointed cusps . . . . Chiroptera

   Size larger, length always greater than 25 mm; no gap between upper incisors; jaw teeth without high, pointed cusps, modified for both grinding and shearing. . . . Carnivora

4. One pair of upper incisors; no more than 4 (pairs of) grinding teeth (molars and premolars) in lower jaw . . . . Rodentia

   Two pairs of upper incisors; 5 pairs of grinding teeth in lower jaw . . . . . . Lagomorpha

After you have identified the skull to its proper order, compare it to the diagrams of Figure 15-2. Label the diagram which corresponds to the order of the skull you have identified by keying. Continue to identify skulls until all of the diagrams of Figure 15-2 are labeled with their proper order. More detailed keys are available in the lab for your use if you wish to identify skulls to correct species.

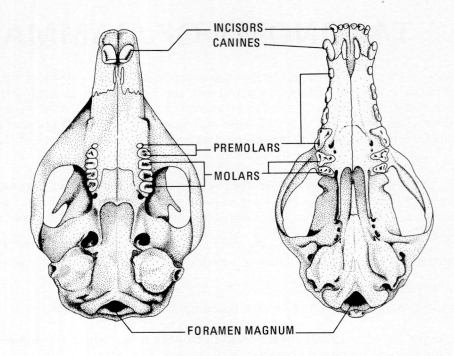

FIGURE 15-1 DIAGRAM OF MAMMAL SKULLS

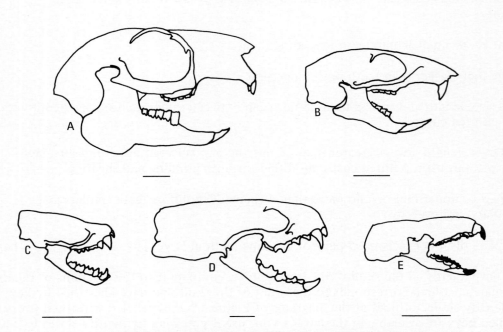

FIGURE 15-2 Mammal skulls of five different orders. Label each drawing with its correct order. Scale lines below each skull represent 1 cm.

A. Order _____   D. Order _____

B. Order _____   E. Order _____

C. Order _____

118

QUESTIONS

1. What is the difference between a molar and a premolar?

2. What is meant by the term **tooth formula**?

3. What is the tooth formula for humans?

4. Which of the orders of mammals you have studied is the most primitive? What are the reasons for your choice?

16. HUMAN GENETICS

Humans are subject to the same laws of heredity as other species of animals and plants. Therefore, it is appropriate to illustrate certain genetic concepts with human examples, which at the same time illustrate the great amount of variability in the human population. Because breeding experiments are not possible, most information about human heredity is derived from twin studies, the analysis of family pedigrees, statistical studies of populations, and, more recently, the study of human somatic cells in cell culture.

This exercise will acquaint you with a few of the relatively common, easily determined hereditary differences in humans. In some cases the genetics of the trait are rather simple; in others, environmental effects may influence the expression of the trait. As the number of traits studied is increased, it should become possible to characterize each individual in the class by a unique "genetic portrait," which is as distinctive as a photograph and usually does not change with time.

This exercise is best carried out as a group exercise. The instructor will discuss each trait, and each student will determine his or her own phenotype for all the traits. Then each student will record his or her phenotype in a table on the blackboard from which the totals and frequencies for the class can be determined.

Test for the individuality of different members of the class by having the entire class stand. Then, as one student calls out his or her phenotype for each trait, students differing from that student for that trait should sit until eventually the student will be standing alone (unless he or she has an identical twin). Through testing various students in turn, it should become evident that each member of the class has a unique combination of traits.

1. **Phenylthiocarbamide (PTC)** is a harmless substance that some individuals find very bitter and others do not taste at all. Nontasters are homozygous recessive (tt); tasters are TT or Tt.

   a. How would you estimate the frequency of heterozygotes among the tasters?

   b. What are the frequencies of the T and t alleles?

2. **Middigital hair** is inherited as a dominant to the complete absence of hair from the middle phalanx of the fingers.

3. **Color blindness** exists in a number of forms, all of them inherited as sex-linked or X-linked traits except total color blindness, which is inherited as an autosomal recessive trait. Because most forms of color blindness are sex linked, the condition is more frequent in males than in females with the frequency of color-blind males ranging as high as 8% in some human populations. Using the charts available, determine whether or not you are color blind, and, if possible, the type of color blindness.

   a. Estimate the frequency of the gene for color blindness in the class.

   b. How would you estimate the expected frequency of color-blind females?

   c. What proportion of the females in the class would be expected to be carriers of the gene for color blindness?

4. **Baldness** is a sex-influenced trait, that is, it is inherited as an autosomal dominant in males and an autosomal recessive in females. Therefore, heterozygous males become bald, but heterozygous females do not even though they can transmit the gene. This difference in expression between male and female heterozygotes is related to the hormonal differences between the sexes. If a heterozygous woman develops a masculinizing adrenal tumor, she too may become bald. Thus the expression of the gene is dependent on the environment, in this case the internal environment. (Most members of the class may be too young for this trait to be expressed.)

5. **Short index finger** is also a sex-influenced trait. Place your hand flat on a table with the fourth or ring finger touching a line at right angles to the length of the finger. Then note whether the second or index finger is on or above the line or is short of the line. The short index finger results from a sex-influenced gene that is dominant in men and recessive in women.

   The inheritance of the traits to be discussed now is less clear-cut than for the traits already treated, primarily because environmental influences, external as well as internal, may prevent the trait from being expressed even though the appropriate genotype is present. This failure is called a lack of **penetrance**. Penetrance is measured as the proportion of affected individuals over the total with the appropriate genotype and may be quite high, approaching 100%, or quite low.

6. **Widow's peak**, in which the hairline forms a distinct point in the center of the forehead, is thought to be inherited as an autosomal dominant. (If you are already bald, consult your baby pictures.)

7. **Hair whorl** on the crown of the head is thought to be genetically determined, with a clockwise whorl dominant to counterclockwise.

8. **Darwin's points** are noticeable tubercles on the infolded edge of the outer rim of the ear, thought to be inherited as a dominant trait. Darwin considered them to be vestiges of the more pointed ears of our primitive ancestors.

9. **Attached earlobes** are considered to be recessive to free earlobes, but neither the trait nor its mode of inheritance are as clear-cut as we might wish.

10. **Tongue curling** was first reported as an autosomal dominant trait, but the discovery of identical twins discordant (different) for the trait cast this interpretation in doubt. There is little doubt, however, that some people can perform more remarkable feats with their tongue than others. In addition to tongue curling (also called tongue rolling), in which the extended tongue is rolled in a U-shape, some people can fold up the tip of the tongue, an ability inherited as an autosomal recessive and independent of tongue curling. Even more remarkable and rare than **tongue folding** is **clover-leaf tongue**, in which the tip of the tongue is formed into a three-leaf clover configuration.

11. **Lateral dominance** can be assessed in various ways.

    a. **Handedness** is not a simply inherited trait and is subject to considerable environmental influence but does nonetheless show some genetic influence as well. A relatively unbiased test of handedness, especially for boys, is to define the preferred hand as the one used to guide the thread in threading a needle.

    b. The **dominant eye** can be determined by sighting with both eyes open on a distant object through a 1/4-inch hole in a large sheet of paper held at arm's length. If you then close each eye in turn, you will find that the object disappears from sight when your dominant eye is closed. Only one eye, your dominant eye, was used to sight the object through the hole in the paper.

Test to determine whether lateral dominance of eye and hand are associated or independent.

12. **Facial dimples** in the cheeks are apparently inherited as an irregular dominant. Perhaps it is appropriate to end your list of hereditary human traits with a smile, for the same muscles we use for dazzling smiles or simpering smirks were used by our aquatic ancestors to aerate their gills.

QUESTIONS

1. If a trait is recessive, must it necessarily be rare in a population? Why?

2. Why do the methods used to study human genetics differ from those used to study other organisms?

3. Why is color blindness more common in men than in women?